Lecture Notes in Economics and Mathematical Systems

483

Springer
Berlin
Heidelberg
New York
Barcelona
Hong Kong
London
Milan
Paris
Singapore
Tokyo

Manfred Königstein

Equity, Efficiency and Evolutionary Stability in Bargaining Games with Joint Production

Springer

Author

Dr. Manfred Königstein
Humboldt-University at Berlin
Institut für Wirtschaftstheorie III
Spandauer Straße 1
10178 Berlin, Germany

Library of Congress Cataloging-in-Publication Data

Königstein, Manfred, 1963-
Equity, efficiency and evolutionary stability in bargaining games with joint production / Manfred Königstein.
p. cm. -- (Lecture notes in economics and mathematical systems, ISSN 0075-8442 ; 483)
Includes bibliographical references.
ISBN-13: 978-3-540-66955-5
1. Game theory. 2. Noncooperative games (Mathematics) 3. Negotiation. 4. Equity. I. Title. II. Series.

HB144 .K66 2000
658.4'0353--dc21

99-087046

ISSN 0075-8442
ISBN-13: 978-3-540-66955-5 e-ISBN-13: 978-3-642-45782-1
DOI: 10.1007/978-3-642-45782-1

"Gedruckt mit Unterstützung der Deutschen Forschungsgemeinschaft D11"

Typesetting: Camera ready by author
Printed on acid-free paper SPIN: 10734499 42/3143/du 543210

Für Birgitta und Moritz
und meine Eltern, Christel und Ewald

ACKNOWLEDGMENTS

First of all, I want to express my deepest gratitude to Werner Güth, my teacher and advisor, for inspiring suggestions, helpful comments, his encouragement and his support whenever needed. Furthermore, I want to thank him as well as Anita Gantner and Reinhard Tietz, as my co-authors on parts of this book. Some groundwork was done during one year of studies at the University of Iowa. I am indebted to Forrest Nelson, for his advice and hospitality. For comments and suggestions I am grateful to Ulrike Graßhoff, Steffen Huck, John Kagel, Ulrich Kamecke, Helmut Lütkepohl, Peter Ockenfels, Stefan Profit, Reinhard Selten, Rolf Tschernig, an anonymous referee and session participants of the Transcoop Workshop '93 by the *Humboldt–Stiftung* (Frankfurt), the *Econometric Society European Meeting* '94 (Maastricht), the IAREP conference '94 (Rotterdam), and participants of brown–bag seminars at the Humboldt–University. The experimental software was developed by Arwed Crüger and Olaf Wellm. My former colleagues at the University of Frankfurt, Gregor Brüggelambert, Christiane Goslar, Frank Heyder, Sigrid Manthey, Robert Pein, Christine Syha and Markus Wendel, helped in running the experiments. My colleagues at the Humboldt–University, Vital Anderhub, Jeanette Bönisch, Tobias Händel, Sabine Kröger, Wieland Müller, Alexandra Schuhmacher, Martin Strobel and Georg von Weizsäcker, provided a nice working atmosphere. Finally, I want to gratefully acknowledge financial support by the *Deutsche Forschungsgemeinschaft*, projects "Entscheidungsvorbereitung und Eingeschränkte Rationalität" and "Strategisches Handeln" as well as SFB 373.

This work was accepted as doctoral dissertation (colloquium at 4th december 1997) by the Wirtschaftswissenschaftliche Fakultät of the Humboldt-Universität zu Berlin.

Berlin, September 1999 Manfred Königstein

Contents

An Overview

The book comprises five essays, each of which is self–contained. I will shortly describe the content of each chapter and discuss their relations.

Chapter 1 is joint work with Reinhard Tietz. It reports an experiment on the ultimatum game with advance production. Two players had to distribute a jointly produced resource via an ultimatum game. The social production function is asymmetric: it exhibits constant marginal productivities that differ between players. Cost are convex in the chosen production quantity. Thus, efficient production decisions are in the interior of the action space. The subgame perfect equilibrium is inefficient and it turns out that the game theoretic solution fits the data quite badly. Contrary, regression analysis reveals that equity theory does explain the data well, on the aggregate level. On the individual level, there is heterogeneity of equity types. A hit rate analysis identifies the empirically most successful equity concepts.

Chapter 2 addresses several problems that arise in statistical analysis of experimental data. Specifically, if subjects interact with one another and are observed repeatedly, measuring the effect of experimental treatments is problematic. The problems are characterized and solutions are suggested. Firstly, due to repeated measurement and strategic interaction, individual decisions may be correlated. It is suggested to consider the data as a panel with strategic interaction groups as units of analysis, and to use modelling to deal with repeated measurement. Secondly, in fixed effect regression analysis, the choice of identifying restrictions imposed on the fixed effect coefficients determines the interpretation of estimation results. A set of restrictions that seems useful for experimental purposes is suggested. Finally, in order to estimate such a model with OLS and to directly produce the coefficients of interest, it is suggested to apply "nested effect coding" of matching group dummy variables.

Chapter 3 is an experimental study on the repeated ultimatum game with advance production. It is the repeated game version of the experiment reported in chapter 1. It investigates subjects' responsiveness to two treatments: repeatedly facing the same opponent ("partnership") and experience. The analysis shows that with experience, behavior converges toward equitable play. A similar effect is induced by partnership compared to one–shot interaction. Furthermore, equity perceptions are heterogeneous and with experience, empirically successful equity concepts even gain in predictive power, while empirically weak concepts, including game theory, become even less important.

Chapter 4 is joint work with Anita Gantner and Werner Güth. It is an experimental study on two simple bargaining games with advance production: the ultimatum game and a symmetric demand game. Compared to the games studied in chapters 1 and 3, social production is simpler, here: cost is linear in the chosen production quantity. The ultimatum game is played "under the veil of ignorance"; i.e. players submit a strategy for both potential roles and the role assignment is determined afterwards. The symmetric demand game is a novel experimental implementation of a game studied by Nash (1950). Analyzing demand decisions as well as acceptance decisions, it is determined which equity standards are more important than others. Since the subjects' decisions are vector choices, the analysis proceeds to classify the vectors. A behavioral model is proposed — "equity anchoring" — which generates testable predictions for the games studied and which is supported by the data. In contrast, game theory explains the observed decisions rather poorly.

Chapter 5 is a theoretical study. Here, an ultimatum game with subsequent production is the base game of an indirect evolutionary game. Two players bargain on a work contract and engage in joint production afterwards. Agents are assumed as being rational with respect to given preferences. Individual effort relies upon the terms of the contract. The subgame perfect equilibrium depends on the production technology and may be inefficient, if the players are egoistic. In contrast, social preferences (modelled as altruism) may lead to an efficient solution. A crucial question is, whether such preferences can be evolutionarily stable, if the reproductive success of an individual depends solely on income. The main result is that they are indeed evolutionarily stable and superior to egoism for generic subsets of the technology set. More specifically, there are subsets in the technology space were only egoism is evolutionarily stable; and for some subsets, both, egoism and social concern are evolutionarily stable. But, most importantly, there exist subsets with positive measure relative to the technology space, for which only social preferences are evolutionarily stable. Accordingly, this result proves that altruism can survive in the long run even in a world where reproduction of preferences solely depends on the monetary payoff received by the inviduals who are endowed with these preferences. Corresponding to the evolution of social preferences the model characterizes the evolution of prosocial behavior; i.e., a positive form of social behavior.

So, three of the five essays[1] report experimental evidence for equitable play in bargaining games with joint production (Chapters 1, 3 and 4). In all three studies, questions of statistical analysis arise, which are generally addressed in Chapter 2. In fact, the regression analysis in chapter 1 is an example of applying the procedures suggested in Chapter 2.

That equity theory has some explanatory power in non–cooperative bargaining games was found out in other experimental studies before. However, including production and allowing for asymmetries in individual contributions to the social resource is more general, and also more realistic. Outside the laboratory, most economic distribution conflicts seem to include some kind of production. It seems to me that distribution conflicts are, basically, conflicts between different perceptions of what is a fair output share given an individual's input share. Consequently, investigating laboratory bargaining with production is important to test the robustness of equitable behavior. So, in the terminology of Davis and Holt (1993) the experiments here should be regarded as *stress tests of equity theory.* A stress test means testing a theory that holds for specific conditions in a broader domain. This aim is different from that of many other studies on bargaining[2] which I want to characterize — relying again on the terminology of Davis and Holt (1993) — as *component tests of game theory.*[3] A component test

[1] As explained above, the essays are self–contained. While this is advantageous for readers that are only interested in parts of the study, those who read it completely will detect some redundancies — for which I apologize. Earlier versions of each chapter appeared as the following discussion papers (see the references): Königstein and Tietz (1998) [Chapter 1], Königstein (1998b) [Chapter 2], Königstein (1998a) [Chapter 3], Gantner, Güth and Königstein (1998) [Chapter 4] and Königstein (1998c) [Chapter 5].

[2] See Roth's description of the development of non–cooperative bargaining experiments (Roth, 1995).

[3] Of course, distinguishing studies according to their research goal is not always as clear as it seems here. Furthermore, many authors may reject to be put into either box. Afterall, the strict separation here is done for the sake of exposition.

is designed to investigate which institutional or environmental components were responsible for the failure of a theory. Many of these studies aim to recover the empirical relevance of game theory.

It turns out here, that in richer environments equity theory yields no unique prediction, which is troublesome in any case. However, the analysis shows that a few equity concepts are empirically superior than others and that each of them is superior to the game theoretic prediction, at least in its simple version. Of course, by adjusting the agents' utility functions many results may be rationalized as equilibrium behavior. But, unless one is willing to impose testable restrictions, such adjustments are meaningless for empirical purposes.

Collecting empirical evidence that is at odds with egoistic, income maximizing, non–cooperative equilibrium behavior may ultimately lead to a revision of standard economic theory. Elements like a concern for equity, altruistic preferences, or other behavioral concepts that seem empirically relevant, may be incorporated into theory or displace previous concepts. For this, it is not enough to simply find counter–evidence to standard theoretical assumptions. It is necessary, in addition, to model such empirical regularities, which might be situation–specific, so that they fit into a more general framework. Furthermore, going from the data toward theory requires to check, whether empirically relevant concepts are theoretically consistent.

Chapter 5 is a study in this spirit. It investigates whether altruistic preferences can be evolutionarily stable, if reproduction is driven solely by monetary success. This seems to me an important consistency check: Why should one believe in altruistic preferences, if they were inferior to egoism in any reasonable model of social evolution?

Even though no preference for equity is modelled directly, which seems natural based on the experimental evidence and remains a theoretical challenge, this study can explain, why moderate, rather than egoistic, bargaining demands occur. It does not allow for disagreement equilibria. However, it is indicated that a stochastic extension of the model — which is left for future research — may yield such equilibria.

1. Profit Sharing in an Asymmetric Bargaining Game

1.1 Introduction

It is a stylized fact of many experiments on the ultimatum game and alternating offer bargaining games that play deviates from the game theoretic solution toward equitable allocations. Evidence for this result are the surveys by Güth and Tietz (1990) and Thaler (1988) as well as the more recent ones by Roth (1995) and Camerer and Thaler (1995).

Nonetheless, there is disagreement regarding explanations of observed behavior. Concerning the explanation of first mover decisions we want to contrast the following three positions:

- game theory
- income maximization and role–asymmetric utility functions
- equity theory

The simple proposition, that subjects are income maximizers and follow a game theoretic equilibrium concept[1] is rejected by most studies. Even in the dictator game[2] and within "double blind" experiments[3] the evidence for this type of behavior is not compelling.[4] Thus, if one thinks that subjects play the game theoretic solution, one needs to allow for more complex utility functions[5], which is difficult to control for in empirical studies. Accordingly, the descriptive value of this view is questionable.

---

[1] An early study that favors such an explanation is Binmore, Shaked and Sutton (1985).

[2] See Forsythe, Horowitz, Savin and Sefton (1994).

[3] Hoffman, McCabe, Shachat and Smith (1994) conclude that subject–experimenter anonymity is important, but this is questioned by Bolton and Zwick (1995).

[4] Some exceptions — i.e., cases were behavior fits the strategic solution reasonably well — are Rapoport, Weg, Felsenthal (1990), Weg and Zwick (1991), Prasnikar and Roth (1992) as well as Harrison and Hirshleifer (1989).

[5] For instance, utility functions that take into account payoff relations between players (see Bolton, 1991).

Roth, Prasnikar, Okuno–Fujiwara and Zamir (1991) found cross–cultural evidence that demands in ultimatum games are income maximizing given the rejection probabilities (which are non–income maximizing). The trouble with this view is that it assumes utilities that exhibit role–asymmetry.

We favor the explanation offered by equity theory:[6] Subjects intend to implement an allocation they consider equitable. There is substantial evidence for equitable sharing in many experiments[7], even though not all observations match this view. We think that the latter is due to heterogeneous player types.

Following this view, our study is designed to reveal heterogeneity of types and, more specifically, heterogeneous equity perceptions. The experiment features an ultimatum game with advance production. The players' productive efforts might stimulate their equity concerns and the game allows for multiple equity standards. For instance, social output might be split proportionally to cost shares or with accounting for the players' asymmetric productivities. So, what constitutes a "fair" allocation is ambiguous.

The game is more general than ultimatum bargaining without production, and we think it is more natural. In reality there is no heavenly manna that only needs to be distributed: Ecomic resources need production, and distribution conflicts concern the relation between inputs and outputs! Nevertheless, it seems that most bargaining experiments focussed on games without production.[8] Advance production can also be viewed as the most natural way to induce entitlement; i.e. the feeling of a bargaining party, that its claim is justified.[9]

Within our study we find a high correlation between input shares and output shares on the aggregate level. Looking at the data on the individual level we, furthermore, find support for heterogeneous equity types. Besides the equal sharing rule, a rather simple notion of equity

[6] See the theoretical work by Homans (1961), Güth (1988, 1994) and Selten (1978).

[7] The occurrence of a high proportion of equal splits was found e.g. in the early study on the ultimatum game by Güth, Schmittberger and Schwarze (1982) and was replicated in various settings many times (see Roth, 1995).

[8] See Roth (1995). However, some related studies with production are Berg, Dickhaut, McCabe (1995), Fehr, Gächter and Kirchsteiger (1997), Gantner, Güth and Königstein (1998, see chapter 4), Güth, Klose, Königstein and Schwalbach (1998), Hackett (1993), Harrison and Hirshleifer (1989) and Königstein (1998a, see chapter 3).

[9] For other attempts to induce entitlement see, for instance, Hoffman and Spitzer (1985) and Güth and Tietz (1986).

(viz. dividing return proportionally to chosen production quantities) exhibit the highest predictive success, with success being determined according to the measure proposed by Selten and Krischker (1983).

1.2 Ultimatum Bargaining with Advance Production

1.2.1 Description of the Game

Suppose two players play a noncooperative game that has two separate phases — production and distribution. Within the production phase the players decide simultaneously upon their individual input quantities q_1 and q_2 causing cost $c(q_1)$ and $c(q_2)$, respectively. Both quantities enter a joint production function and determine a joint output which is to be distributed subsequently according to the rules of the ultimatum game: player 1 states his demand d_1, i.e. the amount of output he asks for himself. Player 2 may accept or reject d_1. In case of acceptance the output is split accordingly, in case of rejection the output is wasted; both players get nothing, but nevertheless they have to bear their respective cost.

So, in this game the "pie" that gets distributed has to be created first via the productive efforts of the players. If the game ends in disagreement, the investments are lost. In more detail and with the parametric restrictions of the experiment, the game is described as follows:

Stage 1: The players choose q_1, respectively q_2, with

$$q_i \in \{0, 1, 2, ..., 30\} .$$

The quantity choices cause individual cost $c(q_i)$ according to the cost table given in the appendix. It lists the cost denoted in German "Pfennig" for each of the discrete production levels and was generated from the following continuous cost function:

$$\tilde{c}(q_i) = \begin{cases} 0 & \text{if } q_i = 0 \\ 15 + 12.5q_i - 1.73q_i^2 + 0.115q_i^3 & \text{otherwise} \end{cases}$$

Thus, the underlying cost function is cubic and the cost values for the discrete production levels were rounded to the next integers. Setting the cost of $q_i = 0$ to zero allowed the subjects not to invest anything at all. Both quantity choices determined the joint return $r(q_1, q_2)$ —

denoted in Pfennig — according to the following social production function:

$$r(q_1, q_2) = 22q_1 + 44q_2 .$$

Social production is asymmetric; player 2 is twice as productive as player 1.

Stage 2: Player 1 is informed about r and chooses his demand d_1, with

$$0 \leq d_1 \leq r$$

and where d_1 could be chosen in increments of 1 Pfennig. Thus, when stating his demand player 1 knows both quantity decisions and the associated cost.

Stage 3: Player 2 is informed about r and d_1 and decides about $\delta_2 \in \{0, 1\}$, with

$$\delta_2 = \begin{cases} 0: & \text{rejection of } d_1 \\ 1: & \text{acceptance of } d_1 . \end{cases}$$

At stage 3, player 2 knows all previous decisions. The description of the game is completed by the payoff functions for either player, Π_1 and Π_2, with

$$\Pi_1 = \delta_2 \cdot d_1 - c(q_1) ,$$
$$\Pi_2 = \delta_2 \cdot (r - d_1) - c(q_2) .$$

So, every player has to bear his individual cost, no matter what the bargaining outcome is.

1.2.2 Efficient Production

Before turning to a strategic analysis of the game we determine both players' efficient production choices. Figure 1.1 displays the marginal return functions as well as the marginal cost function. The abszissa shows the range of possible production quantities 0 to 30. Marginal productivities and marginal cost are measured along the ordinate. Accordingly, marginal returns are greater than marginal cost for low quantities. Maximal surplus $s = r - c(q_1) - c(q_2)$ is reached at $q_1 = 12$ and $q_2 = 16$, the efficient quantity levels. Since the social production function is additive each players net productivity is unaffected by the other player's quantity decision. To grant the efficient actions in the

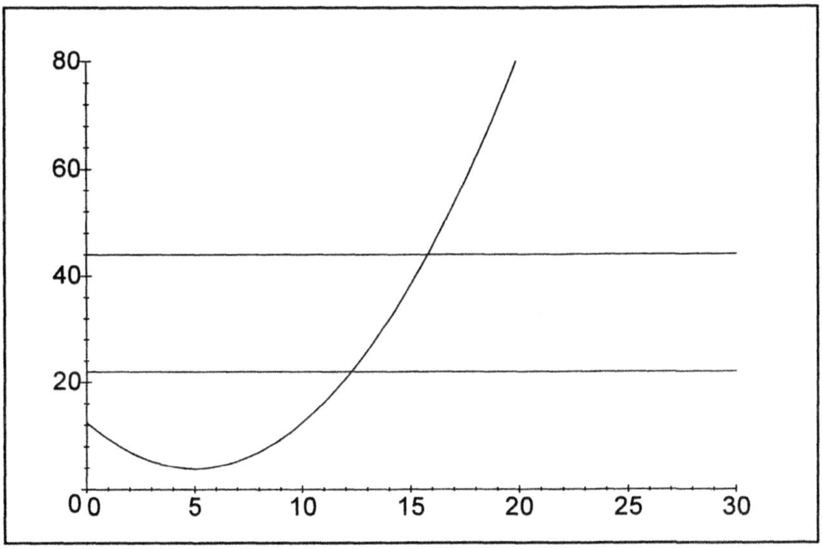

Fig. 1.1. Marginal Productivities of Both Players and Marginal Cost for Different Production Choices

interior of the action space, was intentional for our model specification. It allows for over– as well as under–efficient production, which we think is more natural than corner solutions. The numeric specification was chosen in order to clearly discriminate between the efficient production levels for both players. We expected the experimental decisions to exhibit a considerable dispersion — and it actually turned out so —, such that a good separation between levels that could be chosen on theoretical grounds was desirable.

Given efficient production the return is $r(12, 16) = 968$ and cost are $c(12) = 115$ as well as $c(16) = 243$. Thus, maximal surplus is $s = 610$, and to generate this surplus player 2 has to invest more than twice as much as player 1.

1.2.3 Game Theoretic Solution

A strategic analysis of the game starts at stage 3 and proceeds backwards to the first stage to derive the solution according the notion of subgame perfect equilibrium (SPE).[10] At stage 3 player 2 faces a simple "take-it-or-leave-it" choice. Whenever player 1 leaves nothing but

[10] See Selten (1965) and (1975).

the tiniest "crumb of the pie" for player 2, the latter should choose $\delta_2 = 1$, i.e. accept the demand. Specifically, unless player 1 demands everything, in which case 2 is indifferent between agreement and disagreement, player 2 strictly prefers $\delta_2 = 1$. Accordingly, at stage 2, player 1 should anticipate the result of the third stage and demand (almost) everything. For the production choices (stage 1) this means that player 2 should choose $q_2 = 0$, while 1 should choose his efficient production level $q_1 = 12$. Following these arguments the game theoretic solution is essentially unique and implies the SPE–path:

$$q_1^* = 12 \ , \ q_2^* = 0 \ , \ d_1^* = r - 1 \text{ and } \delta_2^* = 1.$$

The game theoretic solution is inefficient. Player 2 anticipates that he won't get enough to cover his production cost and therefore has no incentive to produce anything at all. Since player 2 is the highly productive player in the game, the inefficiency is considerable. While the efficient surplus was shown to be 6.10 DM, the equilibrium surplus is 1.49 DM. Thus, according to the game theoretic solution player 1 receives a profit of 1.48 DM and player 2 will get 1 Pfennig.

Within this study we are mainly interested in analyzing the demand behavior of subjects. Keeping in mind the game theoretic solution to the distribution phase of the game we will now describe equity theory and discuss its implications.

1.2.4 Equity Theory and Multiple Equity Standards

The substance of equity theory which we present here follows Homans (1961), Selten (1978) and Güth (1994). However, their terminology varies. We follow Güth (1994) and apply the terminology of production theory. He views distribution conflicts as input–output–relationships. We define x_i as individual i's input which enters a social production function defined on the inputs of all members $i = 1, ..., N$ of a society. If y is social output and y_i is the amount of output which individual i receives via distribution, then an equitable allocation is characterized by:

$$\frac{y_i}{x_i} = \frac{y_j}{x_j} \ , \forall i, j \in 1, ..., N. \tag{1.1}$$

Following Güth (1994) we will refer to 1.1 as the "equity norm". It is satisfied if y_i is determined as

$$y_i = \frac{x_i}{\sum_{i=1}^{N} x_i} \cdot y \Longleftrightarrow \frac{y_i}{\sum_{i=1}^{N} y_i} = \frac{x_i}{\sum_{i=1}^{N} x_i} \tag{1.2}$$

Thus, according to 1.2 equity theory simply implies that an individual's output share equals his input share.

While the equity norm is clear cut, the problem of applying it in reality is that the terms "input" and "output" might be ambiguous. E.g. in the game described above it is not clear whether one should view the players' quantity choices as inputs or the associated cost. Actually, it may be this ambiguity which led different authors to use a different terminology for virtually the same substance. Depending on the specifics of the real life problem or theoretical model that is considered one might wish to speak of "investments" and "rewards" — as Homans (1961) — or of a "standard of comparison" and a "standard of distribution" — as Selten (1978).

So, applying equity theory as a solution concept for distribution conflicts may be a problem, if there is no unique way to operationalize it. We will refer to this as the problem of multiple equity standards. It arises if there are multiple candidates to serve as input standards or multiple output standards or both.

Selten (1978) names two requirements that a canditate for an equity standard must fulfill. One is "relevance"; i.e., it must be substantially connected to the problem. The second is "accessibility"; i.e., it must be easily observable by all group members. The latter property excludes hidden variables so that e.g. individual utility may not serve as an output standard.

Within the game we study here, there are several reasonable candidates for input and output standards. Table 1.1 diplays various ways to measure player 1's input and accordingly his input share. Following 1.2 1's output y_1 should equal his input share times total output. In addition to these four input standards there is the "equality standard" which is a potential input standard in every distribution conflict. It claims that each of N members of a society should get $y_i = \frac{1}{N} \cdot y$. For the case of two players this implies a "50–50"-split of the output. In the terminology of production theory it might be called the "personal input standard",[11] but we will refer to it as "equality standard", since this is probably more familiar.

Besides the input side of the equity norm, there is ambiguity in operationalizing the output side as well. We think there are two reasonable candidates for the output standard: return r and surplus s. Combining five candidates for the input standard and two for the out-

[11] See Güth (1994).

Table 1.1. Four Potential Input Standards

Input Standard	Input Share
Quantity: q_1	$\gamma^q \equiv \frac{q_1}{q_1+q_2}$
Cost: $c(q_1)$	$\gamma^c \equiv \frac{c(q_1)}{c(q_1)+c(q_2)}$
Contribution to Return: $22q_1$	$\gamma^r \equiv \frac{22q_1}{22q_1+44q_2}$
Contrib. to Surplus: $22q_1 - c(q_1)$	$\gamma^s \equiv \frac{22q_1-c(q_1)}{22q_1-c(q_1)+44q_2-c(q_2)}$

put standard makes ten different ways for operationalizing equity theory. But, for cost as input standard, both output standards predict the same allocation. Furthermore, the equity standard (γ^r, r) — i.e., "contribution–to–return" as input standard combined with "return" as output standard — is equivalent to (γ^s, s). Thus, we need to consider eight (instead of ten) distinct equity standards.

Some theoretical criteria for the selection among different standards can be found in Güth (1988) and Güth (1994). But here, we will approach the selection problem empirically.

1.3 Experimental Procedures

The experiment was run in October 1993 and January 1994 at the University of Frankfurt/Main with 98 participants altogether — 8 sessions with 12 to 14 subjects each. The participants were mostly students of economics and business administration. They were recruited by handing out information sheets in public places in the economics building and asking them to sign up. Each subject participated in one session only. After taking their seats in the laboratory the participants received written instructions including their personal subject code which they needed later for collecting their earnings.[12] They were divided into two goups, called "players A" and "players B", of equal size and were privately informed about their respective roles. The instructions explained the rules of the game and that each subject would

[12] All instructions are given in the appendix.

play 6 games against changing opponents, but keeping the same role for the entire session. The game rules were read aloud to make them common knowledge.

Decisions were edited into personal computers which were connected via a network. There was no time limit for the decisions. Nobody got to know his opponents' names nor their subject codes. Each subject of group A was sequentially paired with another subject of group B — i.e. we applied a one–population two–type sequential pairwise matching protocol.

Mean earnings for each player were about 16 DM (about 9.40 $ at the time of the experiment) for a 1.5 to 2 hours session including a 5 DM show–up fee. In case of efficient play subjects would have earned 23.30 DM. Earnings according to the subgame perfect equilibrium had been about 14 DM for player 1 and virtually not more than the 5 DM show-up for player 2. Since the subjects could invest money we had to make sure they would pay for potential losses. So, losses in one game were substracted from the show–up fee and the wins of other games and the subjects were informed about this rule. Thus, as long as the sum of all game payoffs and the show–up fee was positive a subject did, indeed, cover all losses it incurred in some of the 6 games. In case this sum was negative, of course, we had no legal way to enforce payment. Only one subject had negative total earnings.

In some sessions the subjects were exposed to different kinds of experimental treatments which we refer to as "decision preparation": a pre–experimental questionnaire and/or an auction for particpation rights.[13] In case of the questionnaire the subjects had to answer several questions regarding their intended choices as well as their expectations for their opponent's actions and their payoffs. There were two kinds of questionnaires: a long "paper questionnaire" with about 20 questions to be answered before game 1 and a short "computer questionnaire" with only three questions to be answered before games 2 to 6.

In case of the auction treatment, after the subjects had read through the game instructions, they had to bid for the right to actually play the 6 games.[14] For each role A and B we had invited 9 subjects, but via the auction we selected only 6 of them to stay for the whole session. Those who did not get a participation right had

[13] The questionnaire and the instructions to the auction are available from the authors upon request.

[14] For an other example of a participation auction in ultimatum game experiments see Güth and Tietz (1986).

to leave after the auction. The auction was organized as a competitive sealed–bid auction[15] — i.e. each of the 6 highest bidders (auction winners) had to pay a price equal to the 7th–highest bid. Subjects received written instructions regarding the auction rules and played one trial–auction for a different object. After auctioning the participation rights, both role prices were made common knowledge.[16]

The questionnaire treatment was applied in 3 sessions, the auction treatment in 1 and both treatments together in 2 sessions. The other 2 sessions (out of a total of 8) featured no preparation treatment at all. The preparation treatments were applied to investigate whether the two methods exert systematic effects on laboratory behavior. In this respect our study is part of a broader investigation of the effectiveness of the two methods within different experiments.[17]

At the end of each session the subjects had to fill out a post–experimental questionnaire; the 16–PA personality questionnaire of Brandstätter (1988). It was used to collect data on individual characteristics of subjects. [18]

1.4 Description of the Decision Data

1.4.1 Production Choices and Acceptance Decisions

The raw data are comprised of the four decisions: the production choices of both players, the demand of player 1 as well as the acceptance decision of player 2 observed in 294 games.[19] As reported in

[15] See Davis and Holt (1993).

[16] In principle the paid role price could effect behavior, since it might strengthen or weaken the asymmetry of the game. The data analyis below actually takes care of this aspect. The regression analyis allows for additive session effects which could reflect differences in the role prices; and the hit rate analysis is conservative: if the role prices had an effect, the observed hit rates would be reduced. Some players might even have tried to implement equity accounting for the difference in role prices. Since individual behavior varies substantially, it seems unlikely that many subjects did so, but afterall we admit that there are further aspects that might improve the empirical relevance of equity theory, which we do not consider here.

[17] For more on the preparation treatments and their effects see Güth and Huck (1997), Güth, Huck and Ockenfels (1996) as well as Königstein (1996b). For a detailed analyis of the answers to the questionnaire and their relation to the subjects' decisions in game 1 see Königstein (1996a).

[18] The influence of personality on ultimatum game behavior is investigated in Brandstätter and Königstein (1998). The questionnaire is contained in the appendix.

[19] Six games by 49 subject pairs.

Table 1.2. Number of Games and Agreements as well as Mean Return, Cost and Profits Per Game

Games	294
Agreements	252 (86%)
Return	8.99 DM
Cost	
Player 1	1.93 DM
Player 2	2.36 DM
Profit	
Player 1	2.07 DM
Player 2	1.57 DM

table 1.2 about 14 % of all games resulted in disagreements; i.e. cases were player 1's demand was rejected by player 2. The average return for all games was 8.99 DM. In games that resulted in disagreement the average return — i.e. the amount of money that both players could have shared among each other — was 7.45 DM. In disagreement games the average amount of money offered to player 2 was 2.23 DM. In addition, of course, player 2 had to bear his cost, but these were sunk cost at the time he decided upon acceptance or rejection. On average, players 1 invested less, but received higher profits than players 2. Thus, players 1 exploited their "first mover advantage" to some degree.

Figures 1.2 and 1.3 display frequency distributions for the quantity decisions for both players. Player 1's efficient production choice is $q_1 = 12$ and one can see in figure 1.2 that this is actually the modal decision. It was chosen in 136 out of 294 cases (46%). For player 2 the efficient production level is $q_2 = 16$ and this is the modal decision of players 2 (76 cases, 26%). Player 2's SPE production is $q_2^* = 0$ which is observed in only 12 cases (4%). Obviously, efficient production was far more attractive to player 2 than equilibrium production.

Besides, there is considerable dispersion. For both roles we observe over–efficient as well as under–efficient production. Risk–aversion could be an ad–hoc explanation for under–efficiency; but, risk attitudes can hardly explain over–efficiency. Rather one should acknowledge the non–linear cost schedule which might have caused difficulties for some subjects. Nonetheless, the distributions are systematically biased toward the efficient quantities.

1.4.2 Demanded Shares of Return and Surplus

Figure 1.4 displays the distributions of demanded return shares. The

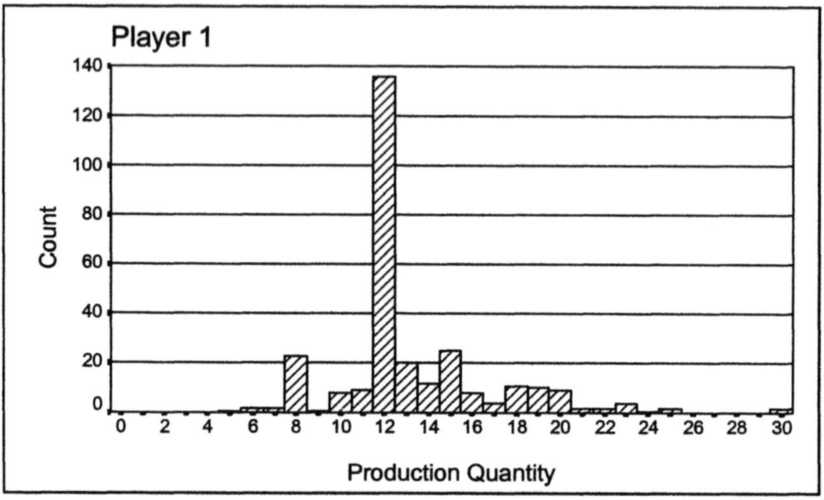

Fig. 1.2. Production Choices of Players 1

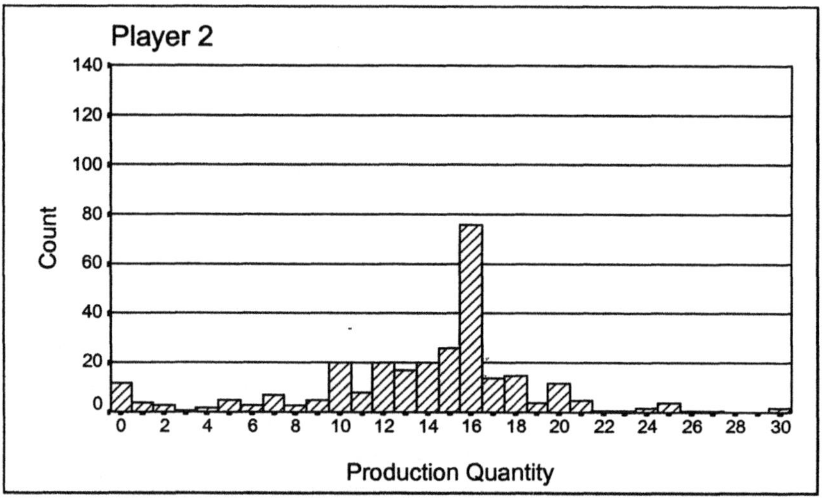

Fig. 1.3. Production Choices of Players 2

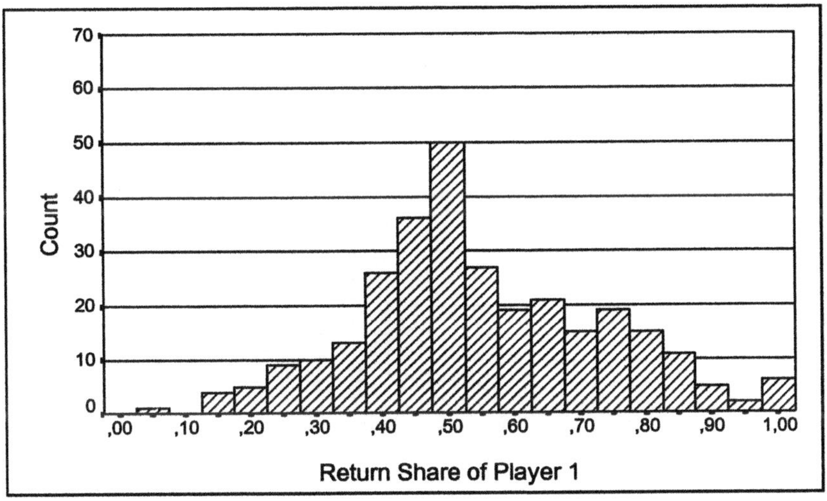

Fig. 1.4. Distribution of Demanded Return Shares

mode is at the "fifty–fifty"–demand, the mean is 0.54. The occurence of many equal splits is similar to what is usually observed in ultimatum games.[20] Also, in accordance with other experiments, but in contrast to game theory the number of extremely high demands is rather low. Demanded shares greater than 90% occurred in only 3% of the cases. Surprisingly, there are considerably many demands below 50 percent of the return (43% of the cases). Such demands rarely occured in ultimatum games without advance production. It seems that it does indeed matter whether the "pie" was produced by the subjects themselves rather than given to them for free as in other experiments.

Considering equity one might wish to look at demanded surplus shares rather than return shares. Namely, a player who demands a high return share is not necessarily greedy, since he may have to bear a high cost share. Also, a pure egalitarian view would certainly claim for an equal split of the surplus. And finally, one might even come up with what we want to call the "modified strategic prediction", which proposes a strategic split of the surplus. Player 1 might expect a rather high risk of disagreement, if he leaves player 2 less than his cost. But

[20] See Güth, Schmittberger and Schwarze (1982) and Forsythe, Horowitz, Savin and Sefton (1994).

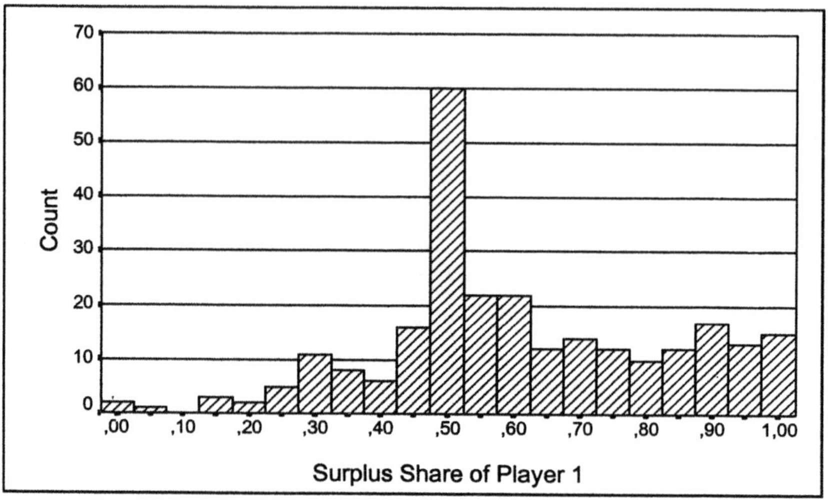

Fig. 1.5. Distribution of Demanded Surplus Shares

granting 2's cost, player 1 may demand the rest for himself — i.e. he demands a surplus share of 1.[21]

Figure 1.5 presents the distribution of demanded surplus shares; they result from dividing player 1's net demand (demand net of cost) by surplus (return minus sum of cost). Only 268 of the 294 cases were considered valid observations on the demanded surplus share. The 26 invalid observations where such that the individual contribution to the surplus was negative for one or both players. These cases are difficult to interprete and were therefore left out.[22] Also, the histogram displays only demanded shares within the unit interval and thus excludes 5

[21] One may argue that, in essence, the "modified strategic prediction" relies on considerations of equity as well, since such considerations are the likely reason for expecting a high risk of disagreement in case a player's cost are not met. However, the reasoning of this concept resembles game theoretic reasoning such that the term seems justified.

[22] An individual contribution to surplus is negative if $q_1 > 19$ or $q_2 > 25$. Suppose e.g. that total surplus and player 1's individual contribution to surplus are negative and he demands more than his own cost. We would probably consider this an expression of greediness. However, calculating the demanded surplus share would give us a negative number, which suggests just the opposite of greediness. Consequently, to avoid confusion we left out these data here.

outliers — 4 cases where the demanded surplus share was smaller than zero and 1 case greater than 1.[23]

One can see in figure 1.5 that the modal class is around the equal split, which supports the egalitarian view. It attracts 22% of the valid cases (60 out of 268). The median is 0.55 and the mean is 0.61. Only 36 demands (13%) ask for a share greater than 0.9 giving weak support for the modified strategic prediction. Notice, that figure 1.5 allows a direct comparison of our results with those of ultimatum games without production. In the latter, demanded shares below 0.5 were rarely observed, but here we find them in 24% of all cases. Our findings, so far, are summarized as follows:

Result 1.1 *Despite the difficult cost function with efficient actions in the interior of the action space, many subjects chose efficient production decisions.*

Result 1.2 *A considerable number of games ended in disagreement. The rejection rate was about* 14%, *replicating the results of ultimatum games without production.*

Result 1.3 *The demands show only little support for strategic behavior. Neither the game theoretic prediction: "Take the whole return!", nor the modified strategic prediction: "Take the whole surplus!", attracted many observations.*

Result 1.4 *The distributions of demanded return shares as well as the demanded surplus shares are considerably dispersed and are different from those observed in ultimatum games without production.*

Result 1.5 *Equal splits are observed quite often, especially equal splits of the surplus, which is evidence for the egalitarian solution of distribution conflicts.*

1.5 Aggregate Demand Behavior

1.5.1 Regression Model

The description of the decision data already indicated that demands might be influenced by production. Furthermore, we know that sub-

[23] In 2 of the 4 cases exhibiting a negative demanded share, players 2 had invested nothing. Thus, players 1 might have feared a high risk of rejection and therefore asked for less than their cost.

jects played a sequence of 6 games and were exposed to different treatment of decision preparation. Thus, we might expect an experience effect and a treatment effect. To allow for the various influencing factors we applied regression analysis. We aggregated the data to yield strategically independent time series and then applied panel data analysis. This procedure was suggested by Königstein (1998b)[24].

Specifically, the subjects were sequentially matched with one another — each player faced 6 opponents —, such that all subjects of one session formed a single interaction group; we will refer to one such group as matching group i. To avoid correlated errors due to strategic interaction we computed mean decisions by matching group i in period t. Thus, instead of 294 individual data points we analyze 8 time series (matching groups) of length 6 (games), i.e. a panel with 48 data points.

The matching groups were exposed to different conditions of decision preparation. Groups 1 and 2 were reference groups receiving none of the treatments. Groups 3, 4 and 5 faced the pre-experimental questionnaire. Group 6 was exposed to the auction treatment and groups 7 and 8 to both treatments.[25] We estimated the following fixed effects model:

$$\begin{aligned}\theta^r_{it} &= \beta_0 + \beta_1\gamma^c_{it} + \sum_{j=2}^{6}\tau_j D^{time}_{jit} + \sum_{j=2}^{4}\lambda_j D^{treat}_{jit} \\ &\quad + \mu_2 D^{group}_{2it} + \mu_4 D^{group}_{4it} + \mu_5 D^{group}_{5it} + \mu_8 D^{group}_{8it} + \varepsilon_{it}\end{aligned}$$

with the variables:

[24] See chapter 2.

[25] The matching groups are nested within the treatment conditions. This requires some concern and will be discussed below.

θ^r_{it} — mean return share demanded by players 1 of matching group i in game t

γ^c_{it} — mean cost share by players 1 of group i in t

D^{time}_{jit} — dummy variables for games 2 to 6, with

$$D^{time}_{jit} = \begin{cases} 1 \text{ if } j = t \\ 0 \text{ otherwise} \end{cases}$$

D^{treat}_{jit} — dummy variables for treatments 2 (questionnaire), 3 (auction) and 4 (both), with

$$D^{treat}_{jit} = \begin{cases} 1 \text{ if group } i \text{ was exposed to treatment } j \\ 0 \text{ otherwise} \end{cases}$$

D^{group}_{jit} — dummy variables for matching groups 2, 4, 5 and 8, with

$$D^{group}_{2it} = \begin{cases} -1 \text{ if } i = 1 \\ +1 \text{ if } i = 2 \\ \;\;0 \;\; \text{otherwise} \end{cases}$$

$$D^{group}_{4it} = \begin{cases} -1 \text{ if } i = 3 \\ +1 \text{ if } i = 4 \\ \;\;0 \;\; \text{otherwise} \end{cases}$$

$$D^{group}_{5it} = \begin{cases} -1 \text{ if } i = 3 \\ +1 \text{ if } i = 5 \\ \;\;0 \;\; \text{otherwise} \end{cases}$$

$$D^{group}_{8it} = \begin{cases} -1 \text{ if } i = 7 \\ +1 \text{ if } i = 8 \\ \;\;0 \;\; \text{otherwise} \end{cases}$$

ε_{it} — error term

and where the coefficients measure:

β_0 — global constant term
β_1 — influence of the cost share
τ_j — fixed effect for game j
λ_j — fixed effect for treatment j
μ_j — fixed effect for matching group j

According to the model the influence of player i's cost share on his demanded return share is estimated while controlling for additive effects

Table 1.3. Regression of the Demanded Return Share on Cost Share, Time, Experimental Treatments and Matching Groups

Regressor	Coefficient		Std. Error	T-Value
β_0	0.316	***	0.075	4.24
β_1	0.443	***	0.154	2.88
τ_2	0.048	*	0.029	1.70
τ_3	0.075	**	0.029	2.61
τ_4	0.101	***	0.029	3.47
τ_5	0.085	***	0.029	2.98
τ_6	0.085	***	0.030	2.84
λ_2	−0.071	***	0.021	−3.41
λ_3	0.067	**	0.029	2.27
λ_4	−0.082	***	0.024	−3.35
μ_2	−0.086	***	0.018	−4.66
μ_4	0.023		0.021	1.11
μ_5	−0.043	**	0.019	−2.30
μ_8	0.029	*	0.017	1.76
Model				
F	12.12	***		
$R^2_{adj.}$	0.76			

of time, treatment and matching group.[26] The estimation results are shown in table 1.3.[27] The model is highly significant and explains 76% of the variance.

1.5.2 Influence of Cost

The influence of the cost share is positive — which it should be following equity theory. It is economically substantial and statistically significant on the 1% level. This clearly confirms the importance of equity considerations for bargaining behavior and rejects the subgame perfect equilibrium; namely, the latter implies $(\beta_0 = 1, \beta_1 = 0)$.

If all subjects exhibited equity type (γ^c, r) as proposed by the model, then the coefficients should be $(\beta_0 = 0, \beta_1 = 1)$ which is rejected as well. This is not too surprising, since there may be heterogeneous equity types. Type (γ^e, r) predicts $(\beta_0 = 0.5, \beta_1 = 0)$. Consequently, if the subject population were a mixture of types (γ^c, r) and

[26] The dependent variable in our regression analysis is bounded above and below, such that the standard assumptions about the error distribution is questionable. But, among aggregated data there were no boundary observations such that our simple model seemed a good approximation.

[27] Superscript stars next to the coefficients indicate significance on the 1–, 5– or 10–percent level.

(γ^e, r), the equity prediction would be roughly $(\beta_0 \in [0, 0.5], \beta_1 \in [0, 1])$ which is consistent with the data.[28]

However, which equity types are present in the population and in which proportions cannot be judged on the basis of the regression analysis. The input standard $\gamma^c, \gamma^q, \gamma^r$ and γ^s and furthermore the output standards r and s are positively correlated — even though not linearly. So, besides (γ^c, r) and (γ^e, r) other equity types may be present. Inclusion of more than one equity standard in the regression model results in multicollinearity. Therefore we refrained from doing so. We tried other models with single equity predictors. The results were qualitatively the same, but there were minor reductions in goodness of fit compared to the result we report here.

1.5.3 Experience

The influence of experience is captured by the dummy variables for games 2 to 6. The coefficients represent the size of additive shifts compared to game 1, which is the reference category.[29] The coefficient 0.048 for game 2 (0.75 for game 3) implies that, other things being equal, the demanded return share was on average 4.8% (7.5%) higher than in game 1. Looking at all game dummies we find that the shift in demand was pronounced in early games and stabilized in later ones; i.e., demand seems to converge with experience. While all differences compared to game 1 are statistically significant, the differences between later games are not.

The increase in the demanded return share could mean that some or all subjects learn to exploit their first mover advantage to some degree. But, one should note that even with experience demand stays well below the game theoretic level. Furthermore, since behavior seems to converge it is unlikely that allowing for more experience leads to demands near the game theoretic solution.

[28] In fact, if there are only two types of players one should expect the estimated aggregate coefficients to be a linear combination of the coefficients for each type. However, since the population may be a mixture of many types, we do not want to make a precise statement here.

[29] Thus, the dummies are indicator–coded and the game 1 dummy was left out in the estimation. Otherwise the model would not be identified due to linear dependence of the time dummies and the unit vector which is used implicitely to estimate the global constant.

1.5.4 Decision Preparation

Applying the questionnaire decreased the demanded share by 7.1% (λ_2) compared the reference category "no preparation". In contrast, the auction led to an increase of 6.7% (λ_3). This is in line with what Güth and Huck (1997) as well as Güth, Huck and Ockenfels (1996) observed. The coefficient (λ_4) indicates some interaction between the treatments when both are applied within the same session. It is not significantly different from (λ_2).

The questionnaire asked the players, among other things, to consider the opposite player's decisions. For instance, they were asked what they expected the respective other player to do and what they would do themselves given the other role. We think that these questions might have reduced the subjects' perception of the strategic asymmetry of the game.

On the other hand, auctioning participation rights might have increased the awareness of the asymmetry. Namely, it forced subjects right at the beginning of the session, to consider the feasible payoffs and to back their expectation by a monetary bid. Another possible explanation for the auction–effect is entitlement: the auction winners have earned the right to exercise strategic power, which subjects feel less entitled to without an auction.[30]

Overall, these findings show that laboratory bargaining behavior is to some degree influenced by framing. One should note that the auction effect works into the same direction as experience — both lead to an increase in the demanded return share —, while the questionnaire effect runs against experience.

1.5.5 Fixed Effects of Matching Groups

The matching group dummies were included in the model to take care of repeated measurement. They are effect–coded within the treatment groups such that the matching group effects sum to zero within each treatment group.[31] The dummies for matching groups 1, 3 and 7 — i.e., the first dummy within each treatment group — were removed from the model to yield identification. The respective coefficients are implicitely set to

---

[30] See Güth and Tietz (1986).

[31] For a more general description of this kind of dummy variable coding and its implications for the interpretation of the regression results see chapter 2.

$$\mu_1 = -\mu_2$$
$$\mu_3 = -(\mu_4 + \mu_5)$$
$$\mu_7 = -\mu_8.$$

Furthermore, the fixed effect of matching group 6, μ_6, is identical to λ_3, since only in session 6 the auction was applied as sole preparatory treatment.[32]

The coefficients display the difference of matching group i from average behavior within the respective treatment group. Some of them are significant indicating that considering repeated measurement and estimation of matching group effects is reasonable. We do not know the causes of these matching group biases, and so do not want to further interpret them.

1.5.6 Summary of Regression Results

We summarize the findings as follows:

Result 1.6 *In the aggregate, the cost share exhibits a substantial and significant influence upon the demanded return share. The evidence rejects game theory and supports equity theory, but there is heterogeneity of equity types.*

Result 1.7 *With experience, demanded shares increase, but stabilize at a level far below the game theoretic one.*

Result 1.8 *Laboratory bargaining behavior is affected by decision preparation. The auction treatment shifts demands upward; the questionnaire treatment shifts demands downward.*

1.6 Individual Demand Behavior

1.6.1 Equity Types as Point Predictors or Area Theories

The regression analysis supports the proposed model as a reasonably good description of aggregate behavior. But it left open whether some subjects use different equity standards than others. In addition, egalitarian or game theoretic reasoning could apply at least to some individuals.

[32] The auction treatment was also applied in sessions 7 and 8, but together with the questionnaire treatment.

If subjects exhibit heterogeneous types, it would be worthwile finding out which types are more important than others. Even though an equity standard or strategic reasoning may not explain the aggregate very well, it might be followed by a considerable proportion of subjects.

Therefore, we will apply a procedure, which we refer to as "hit rate analysis", to determine the predictive power of the different concepts on the basis of the individual decision data. It is based on hit rates of area theories and a measure of predictive success for area theories m, which will be defined below and which was proposed by Selten and Krischker (1983):

- First, we will turn point predictions into area theories by allowing for prediction errors within certain ranges. We will compare the predictive success of different theories by their hit rates for given ranges.
- Secondly, in order to compare area theories for different levels of accuracy, we introduce the notion of "area" a by Selten and Krischker, which is equivalent to what we define as the "range" for theories based on return (as output standard), but which differs from the range for theories based on surplus.
- Finally, we determine for each theory the area a^* that maximizes m subject to certain constraints. We can thus compare the candidate theories by their values m^* which is the (constrained) maximal predictive success of the respective theory.

If a subject demands exactly the output share suggested by a theory, this counts as an exact hit of the respective theory. The proportion of hits among all decisions is the hit rate.

Given an input standard as well as an output standard, equity theory makes a precise prediction "how to cut the pie." But subjects might follow an equity standard only roughly. They might favor round numbers instead of calculating an equitable demand exactly.[33] If the demanded share is in close distance to the equitable share, one may still want to consider it a hit even though not an exact one. Instead of looking for exact hits we will therefore determine hit rates for lower levels of accuracy. E.g. if a player's demanded share lies within a distance of ± 5 percentage points of the share suggested by theory, this will be

[33] A formal description and some experimental evidence of subjects' preferences for numbers that are prominent in the decimal system can be found in Albers and Albers (1983), Albers (1997) as well as Albers, Albers, Albers and Vogt (1997). Their concept of prominent numbers is more specific than rounding. Choosing a prominent number implies rounding, but not vice versa.

Table 1.4. Hit Rates in Predicting the Demanded Return Share for Different Ranges as well as Maximal Success Values and the Corresponding Hit Rates and Areas

Range	Equality	Quantity	Cost	Contr. to Return	Contr. to Surplus	Subgame Perf. Eq'm
1%	6.1	8.2	6.1	4.8	1.5	1.4
2%	8.5	13.3	8.8	4.8	1.5	1.4
5%	16.7	27.6	15.3	7.8	4.5	2.0
10%	29.3	38.1	25.2	11.9	8.2	2.4
a^*	10.0	10.0	10.0	1.0	1.0	1.0
h^*	29.3	38.1	25.2	4.8	1.5	1.4
m^*	19.3	28.1	15.2	3.8	0.5	0.4

considered a hit on a 10% range. So, we extent the point predictors by a tolerated prediction error, such that calculating and comparing hit rates can be done more reasonably.

1.6.2 Hit Rates of Return Share Predictions

The size of the tolerated prediction error is matter of choice. Therefore, we calculated the hit rates of the various predictions for the demanded return share not only for a single range but for several different ranges.

The result is shown in table 1.4. It reports the hit rates (in percent of a total of 294 observations) of predictions based on return as output variable for ranges 1%, 2%, 5% and 10%.[34] Equality, quantity, cost, contribution–to–return and contribution–to–surplus are the input standards we derived above. In addition the hit rates for the subgame perfect equilibrium are shown; i.e. the numbers in this column represent the percentage of demanded shares within the neighborhood of the game theoretic prediction.

One can see, for example, that the quantity standard hits in 13.3% of all cases on a range of 2% — i.e. within ± 1 percentage points of the predicted return share. To illustrate these numbers suppose, e.g., $q_1 = 10$ and $q_2 = 15$ such that the return is 880. The quantity standard predicts that player 1 demands 40% of the return, i.e., $d_1 = 352$. The hit rate says, given this scenario, that actually 39 out of 294 subjects (13.3%) demand shares between 39% and 41% of the return — i.e. demands are within the intervall $[344, 360]$. On the 10% range the quantity standard hits in 38.1% of the cases. For the given example

[34] The meaning of a^*, h^* and m^* will be explained below.

this would mean that almost $\frac{2}{5}$ of all observable demanded shares are between 35% and 45% of the return — accordingly demands are within the intervall [308, 396].

Comparing the hit rates on a range of 10% we find that the quantity standard is the best predictor. The equality standard is second best, and the cost standard is third best. These three are substantially better than all others. The game theoretic prediction does worst of all. One might argue that it predicts a corner solution such that the range should be adjusted, since it is impossible to observe a demanded return share greater than 1. This is a valid point. But even if we compare the entry on the 10% range for the subgame perfect equilibrium (2.4%) with the entries on the 5% range for the other predictors, the result for game theory is worst of all. The first three input standards are superior to the last three for all ranges. Each of the first attracts a considerable share of the observations, while the latter three do not. It should be noted that the predictions are non–exclusive. There is some overlap; e.g., if both players choose the same quantity level, the input standards based on quantity, cost or equality all predict the same demanded share.

1.6.3 Hit Rates of Surplus Share Predictions

Table 1.5 reports the hit rates of predictions based on surplus as output variable.[35] The prediction ranges are now measured in percentage points of the respective surplus. We find on the 10% range that "equality" is predicting best, "quantity" second and "cost" third. The modified strategic prediction and the contribution–to–surplus standard are worst.

The average surplus is only about 52% of the average return. This means that a theory that predicts a division of the surplus within a range of 10% is as accurate on average as a theory that predicts a division of the return within a range of 5.2%. Thus, the hit rates in

[35] We could have excluded the entries for "cost" and "contribution to surplus". We have argued abvove that (γ^c, s) and (γ^c, r) are equivalent — i.e., they imply the same demand — and, furthermore, that (γ^s, s) and (γ^r, r) are equivalent, such that the information is already contained in table 1.4. But, for ease of comparison we report them as well. Namely, predicting an area of e.g. 10% of the available surplus is, in general, more accurate than predicting an area of 10% of the available return, such that the reported hit rates differ between table 1.4 and table 1.5.

Table 1.5. Hit Rates in Predicting the Demanded Surplus Share for Different Ranges as well as Maximal Success Values and the Corresponding Hit Rates and Areas

Range	Equality	Quantity	Cost	Contr. to Return	Contr. to Surplus	Modified Strat. Pred.
1%	12.7	3.7	5.2	1.1	3.4	2.6
2%	16.0	7.5	6.3	1.9	3.7	3.7
5%	22.4	13.1	9.3	7.1	4.9	5.6
10%	30.2	20.9	16.0	10.8	7.5	7.5
a^*	5.8	5.8	5.8	3.5	0.6	1.7
h^*	30.2	20.9	16.0	8.2	3.4	4.9
m^*	24.4	15.1	10.2	4.7	2.8	3.2

table 1.5 on the 10% range can approximately be compared with those in table 1.4 on a 5% range.

However, one can systematically adjust for differences in accuracy by applying the measure of predictive success proposed by Selten and Krischker (1983).[36] Accordingly, the predictive success of area theories m is defined as:

$$m = h - a$$

with the hit rate h and the area a. The area is measured as the predicted outcome space divided by the possible outcome space. Here, for theories based on return, the possible outcome space is the unit interval, since these are the possible shares of return a player can demand. So, the concept of area is equivalent to the range and, therefore, the success measure is simply the hit rate minus the respective range.

For theories based on surplus, the size of the range is not equivalent to the area, since a player can demand a surplus share greater than 1 or smaller than 0. As explained above the surplus was on average only 52% of the amount of return. Thus, we calculated the areas for theories based on surplus by multiplying the repective ranges with the factor 0.52.

Finally, since we calculated hit rates for different ranges, one can calculate the success measure m for different areas as well. We did that and selected the maximal measure m^* as well as the corresponding area a^* and hit rate h^* to be displayed in tables 1.4 and 1.5 (all values in percent).

[36] For a discussion of the measure and a comparison to some alternative measures see Selten (1989).

1.6.4 Comparison and Discussion of Results

Now, we can compare the various concepts by their predictive success determined as the maximal success measure within ranges 1%, 2%, 5% and 10%. Accordingly, the best predictor among all candidates is equity theory based on quantity (as input standard) and return (as output standard). This concept reaches maximal predictive success $m^* = 0.281$ for $a^* = 0.10$. It is one of the most simple concepts. In applying this equity standard, a subject has to do less calculations than e.g. in applying surplus as output standard. Given that economic agents exhibit cognitive limitations such behavior may be reasonable. It is guided by a concern for equity but regards internal cost for the processing of information.

The second best predictor is equity theory based on equality (input) and surplus (output). The combination of equality as input standard and return as output standard is third. So, egalitarian reasoning seems to drive the behavior of a large number of subjects. Purely egalitarian sharing certainly claims for equal surplus splitting. That, nevertheless, equality combined with return does so well might, again, be due to cognitive limitations.

Surprisingly, distributing output proportionally to the players' cost shares has considerably less explanatory power than the equity concepts named above. While cost measures a player's effort in "real" economic terms, the chosen quantity is only a nominal variable. It is positively, but not linearly, correlated with cost. In terms of Selten (1978) one might say that the cost standard is more relevant to the problem, but the quantity standard is more accessible.

The most complex equity standard is the combination of contribution-to-surplus (input) and surplus (output). It takes into account the players' cost as well as their productivities, but its explanatory power is weak.

Furthermore, the analysis allows to compare equity theory with strategic concepts and one can clearly see that the latter explain the data rather poorly. The analysis of individual demand behavior shall be summarized as follows:

Result 1.9 *Forming areas around predicted points and applying the measure of predictive success for area theories according to Selten and Krischker (1983) one finds that equity theory outperforms strategic concepts.*

Result 1.10 *Simple equity standards were superior to more complex ones, even though the latter might be economically more relevant.*

Result 1.11 *Equity based on quantity (input) and return (output) was most successful and received a hit rate of almost $\frac{2}{5}$ within a range of 10%, i.e., within a distance of ±5% of the predicted share.*

Result 1.12 *A substantial share of observations is consistent with egalitarian considerations.*

1.7 Summary

The study investigates the predictive power of equity theory and strategic concepts within ultimatum bargaining with advance production. This is different from many other experimental studies on non-cooperative bargaining games, since usually the ressource to be distributed between the subjects is given by the experimenter. Here, the "pie" is produced via joint production effort. This scenario is more general than bargaining without production, and we think it is more natural as well.

The description of the raw data showed that the production choices are considerably dispersed, but nevertheless systematically biased toward the respective player's efficient production level. The distributions of demanded return shares and demanded surplus shares partly replicated findings of other studies; namely, a low percentage of high demands and a considerable proportion of equal surplus splits. But, they also indicate that "advance production" compared to "no production" should be considered an important treatment in studies on ultimatum bargaining.

Even stronger evidence for the impact of input shares on demanded output shares was found in the regression analysis. Controlling for influences of time, decision preparation and matching groups there was a high positive correlation between cost shares and demanded return shares. Overall, the model explained aggregate behavior quite well. The demanded return share increased over time and seemed to converge to a level far below that proposed by game theoretic reasoning.

Since the regression analysis did not allow for heterogeneous equity types, we applied a hit rate analysis to determine which equity standards were more successful in explaining the data than others. Accordingly, a quite simple concept was the best predictor; namely, taking quantity as input standard and return as output standard. While this equity standard is easily accessible for boundedly rational agents, there are other standards, e.g. based on cost as input standard, which seem economically more relevant but were less successful. Equity based on equality (input) and surplus (output) was also important for a large number of subjects. While this is similar to findings in ultimatum games without production, it should be noted that here an equal split of the surplus is far less obvious, in most cases. So, the evidence supports the robustness of equal surplus splitting as an important sharing rule.

Overall, equity theory can explain a substantial share of observations, whereas only few observations are consistent with purely strategic reasoning. Not only the game theoretic prediction performed badly, but also a less extreme strategic prediction: "Offer player 2 enough to meet his cost and take the rest!" On the basis of our experiment we cannot exclude that subjects follow a mixture between strategic reasoning and equity considerations; but, nonetheless, we show that considerably many subjects follow equity theory rather closely.

2. Measuring Treatment–Effects in Experimental Cross–Sectional Time Series

2.1 Introduction

In the analysis of individual decision data collected in experimental games problems arise due to repeated measurement and strategic interaction of subjects; namely, that the decisions are non–independent. These problems are common to experimental economists. For example, with respect to repeated measurement Roth argues in *The Handbook of Experimental Economics* that "not only autocorrelation, but also potential learning effects (diminishing variance by periods) raise questions that need to be addressed in analyzing the data. There remains considerable room for improvement in econometric methods and tests to address these issues."[1]

The study contributes to this research program by suggesting a procedure for experimental data analysis:

1. Consider experimental data as panel data sets with strategic interaction groups being the units of analysis.
2. Use modelling to deal with repeated measurement.

Given this procedure is followed, the second part of the paper describes a more special, econometric problem that arises in regression analysis of experimental panels, if different groups of subjects are exposed to different experimental treatments. Within a fixed effects model that allows for treatment effects as well as matching group effects different identifying restrictions can be imposed. A set of restrictions that seems quite useful for experimental purposes will be suggested:

- The matching group coefficients should sum to zero within each treatment group such that the matching group effects cancel.

[1] Roth (1995), p. 334, note 26.

- The coefficient for the "zero–treatment" group should be normalized to zero such that the treatment effects are expressed in comparison to benchmark–behavior.

Finally, three different alternatives for estimating the model of interest are described. In case OLS shall be applied, it will be suggested:

- Use nested effect coding for matching group dummy variables to directly produce the required estimates and their statistics.

2.2 Experimental Data, Repeated Measurement and Strategic Interaction

Quite often, the data collected in laboratory experiments are cross–sectional time series. E.g. suppose a one–person experiment in which a subject has to decide upon a dynamic decision problem or several static problems. Thus, one observes a time series of individual decisions and the data of several individuals make a cross–sectional time series data set. When analyzing such data by means of regression analysis one has to consider the "repeated measurement" problem: i.e., the error terms might be correlated, since each decision unit was observed several times. One can deal with this problem by modelling the error structure. Procedures for doing so are presented e.g. in the textbook by Judge et. al (1985).

Data analysis is considerably more problematic, if the individual decisions were collected in interactive games. For example, if two subjects play a repeated game, this generates a sequence of decisions and game outcomes. The data of several pairs comprise a panel data set. Other examples are experiments in which subjects play several static games against changing opponents or dynamic market experiments. In these experiments a few or many subjects interact with one another. Thus an individual's decision at some point in time might be influenced by some other player's decison in the past. One can e.g. easily imagine in an experiment on public goods provision that the willingness to contribute to the public good depends on the contributions of other players in the past. In regression analysis of the individual decision data this means again that the error terms are correlated. I will refer to this problem as the "strategic interaction" problem.[2] The source for

[2] Some problems of analyzing data which are non–independent due to strategic interaction are discussed in Davis and Holt (1993), p. 527–528.

the correlation of errors is the strategic interaction of individuals such that it shall be distinguished from the repeated measurement problem.

There are two extreme ways of dealing with strategic interaction and repeated measurement, and there are examples in the experimental literature for both. One is to simply ignor the potential correlation of errors and to analyze the data as if the errors were uncorrelated. The other one is to take the mean of all decisions within one interaction group. While the first method is rather risky in all cases — and probably wrong in many — the second is extremely conservative and is wasting information; since all decisions of one interaction group are aggregated, the dynamic structure is lost.

There are less extreme ways to deal with the two problems. The way proposed here is to consider the interaction group as unit of analysis, which delivers a single time series, and to apply panel analysis procedures to a set of interaction groups, which take care of repeated measurement. Accordingly, an observation on interaction group i at time t is a single data point. This data point may represent the decision of one, few or all players within group i. E.g. in symmetric games it is natural to take the mean decision at time t of all players in group i. Thus, doing cross–sectional time series analysis on interaction groups as unit of analysis is conservative with regard to strategic interaction, but not with regard to repeated measurement.

One could go further than that and try to model the error structure due to strategic interaction as well. In many experiments the sequencing of decisions is controlled by the experimenter or can be retrieved ex–post. For example, subjects might be sequentially pairwisely matched according to a specific matching protocol. This information could be used to model the error structure. However, it seems rather difficult. The influence of other players' decisions may be non–linear; e.g. it might well be that only extreme kinds of behavior are percieved by other players and enter their decisions making in the future. Also, there could be second–order effects, third–order effects, and so on. To avoid these difficulties the analysis proceeds as proposed above.

In some experiments an interaction group consist of all subjects participating in the same experimental session. This is the case, for example, in dynamic market experiments in which all subjects interact in each period. But, if, for instance, a session group is subdivided into a few matching groups such that strategic interaction takes place

Table 2.1. Partitioning of the Set of Matching Groups According to Experimental Treaments

Treatment Group	Matching Groups
1	$1, ..., N_1$
$\vdots$	$\vdots$
k	$N_{k-1}+1, ..., N_k$
$\vdots$	$\vdots$
K	$N_{K-1}+1, ..., N_K$

only within the matching groups, then each of the subgroups may be considered an interaction group.

Of course one might argue that subjects participating in the same session see each other and that this is a form of interaction in any case. Consequently, one would always have to choose session groups as units of analysis. But, the emphasis here is on the data problems induced by **strategic** interaction — according to the rules of the game studied — as opposed to any kind of interaction. In the sequel, therefore, matching groups will be considered as unit of analysis, each delivering a time series that is **strategically independent** of other matching groups. Nevertheless, it should be mentioned again that there are many ways of dealing with potentially correlated errors between the two extremes named above, and one might want to be more conservative in estimation than what is proposed here.

2.3 Fixed Effects, Identifying Restrictions and Interpretation

Suppose an experiment in which matching groups $i = 1, ..., N$ were observed at times $t = 1, ..., T$. Furthermore, suppose that among the population of matching groups different subpopulations were exposed to different experimental treatments $k = 1, ..., K$ such that the set of matching groups is partitioned as shown in table 2.1. Let y_{it} represent the decision of group i in t which is to be explained and let y represent the according vector of observations. It is natural to expect that y depends on a set of explanatory variables x_m, with $m = 1, ..., M$, as well as on time, treatment, a matching group component and a stochastic error term. Using vector notation a simple specification for these dependencies is given by the following fixed effects model:

$$y = \beta_0 + \sum_{m=1}^{M} \beta_m x_m + \sum_{l=1}^{T} \tau_l d_l^{Time} + \sum_{k=1}^{K} \gamma_k d_k^{Treat} + \sum_{k=1}^{K} \sum_{j=N_{k-1}+1}^{N_k} \mu_j d_j^{Match} + \varepsilon \qquad (2.1)$$

where y, x_m, d_l^{Time}, d_k^{Treat}, d_j^{Match} and ε represent $N \cdot T \times 1$ vectors, with the variables:

d_l^{Time} — dummy variable for period l,

$$d_{lit}^{Time} = \begin{cases} 1 \text{ if } t = l \\ 0 \text{ otherwise} \end{cases}$$

d_k^{Treat} — dummy variable for treatment k,

$$d_{kit}^{Treat} = \begin{cases} 1 \text{ if } N_{k-1} + 1 \leq i \leq N_k \\ \quad \text{(i.e., } i \text{ belongs to treatment group } k\text{)} \\ 0 \text{ otherwise} \end{cases}$$

d_j^{Match} — dummy variable for matching group j,

$$d_{jit}^{Match} = \begin{cases} 1 \text{ if } i = j \\ 0 \text{ otherwise} \end{cases}$$

ε — error term

and the coefficients:

β_0 — global constant term
β_m — influence of regressor x_m
τ_l — fixed effect for period l
γ_k — fixed effect for treatment k
μ_j — fixed effect for matching group j

The model proposes that everything else being equal the decision of matching group i in period t exhibts additive shifts for period t ("time effect"), treatment k ("treatment effect") and matching group i ("matching group effect"). Modelling the influence of time by time dummies, as it is suggested here, allows for a more general time influence compared to e.g. a linear influence of a time index. But, of course there is no general rule how to model the influence of time. Its very specification is actually not an important issue for what is discussed below. However, that one is able to model the time influence somehow is, afterall, an advantage of considering experimental data as cross–sectional time series compared to using aggregates for sessions (matching groups). Therefore, a time influence is included here and in all subsequent models.

Model 2.1 cannot be estimated directly, since there is linear dependence among the regressors deriving from several sources. Firstly, the time dummies, the treatment dummies as well as the matching dummies — i.e., each set of variables separately — add–up to the unit vector x_0 which is implicitly contained in the model to estimate the global constant term. In addition, summing over all matching group dummies within treatment group k results in a vector that equals d_k^{Treat}.

A standard procedure to get identification is to drop enough dummies out of the regression equation. For example, one could eliminate one of the time dummies, one of the treatment dummies and one matching dummy within each treatment group. Formally, this means that the coefficients of the omitted variables are restricted to zero. Suits (1984) has shown that other identifying restrictions can be employed and pointed out that care must be taken with respect to interpretation.

He examined the dependence of household expenditure for gasoline on income and region of residence. Each data point was drawn from one of four regions in the US. Dummy variables were used in the analysis to measure the regional influence. Regarding interpretation two questions were considered:

- What are the dummy coefficients supposed to express?
- What is the intercept term supposed to express?

Restricting one dummy coefficient to zero meant that this region was set as the reference class. It implied that the coefficients estimated for the other three regions expressed the difference in behavior of people living in the respective region compared to those living in the reference region. At the same time it meant that the intercept term of the regression equation picked up the effect of the reference region such that the estimated intercept did not represent US–average behavior. Suits showed that restricting the sum of the coefficients to equal zero was advantageous in his case, since then the intercept term did indeed represent US–average behavior. At the same time it implied that the dummy coefficients represented the difference of people living in a certain region compared to the national average.

Of course, in Suit's case the two questions regarding interpretation can not be solved independent of each other, but for the issue here —

which is a slightly complicated variant of Suit's problem — it is useful to think of them separately.

Namely, in model 2.1 the matching groups are nested within the treatment groups. Accordingly, γ_k is the dummy coefficient for treatment k and at the same time it is a constant term for all matching groups contained in treatment group k. Its interpretation therefore depends not only on the restrictions imposed on the parameters γ_k themselves, but also on the restrictions imposed on the coefficients μ_j for all matching groups contained in treatment group k.

Now, being aware of the problems involved in measuring the treatment effect, one might ask what kind of interpretation is preferred in experimental economics. Of course there is no unique answer, but I think in most cases one would like the matching group effects to cancel within each respective treatment group; i.e., one imposes the restrictions:

$$\sum_{j=N_{k-1}+1}^{N_k} \hat{\mu}_j = 0, \forall k = 1, ..., K. \tag{2.2}$$

Regarding the treatment coefficients one may wish to impose either

$$\sum_{k=1}^{K} \hat{\gamma}_k = 0 \tag{2.3}$$

or

$$\tilde{\gamma}_1 = 0. \tag{2.4}$$

Requiring conditions 2.2 and 2.3 implies that the coefficients $\hat{\gamma}_k$ measure the treatment effect as deviation from average behavior of the subject population. Alternatively, imposing conditions 2.2 and 2.4 means that the coefficients $\tilde{\gamma}_k$ measure the treatment effect as difference to the reference category $k = 1$. In exeriments, quite often one likes to compare behavior induced by a certain treatment with benchmark behavior observed in a "zero–treatment" group. In this case the second alternative is the preferred one.

2.4 Estimation Procedures

2.4.1 Restricted Leased Squares or Transformation of OLS–Coefficients

To produce the required estimates the most elegant procedure is to use Restricted Leased Squares (RLS), whenever it is feasible, and to directly imposed the necessary restrictions on the coefficients. But, not all standard statistics packages offer this option. In addition, Greene and Seaks (1991) argue that in standard textbooks the RLS–estimator is presented in a way "that is lacking in generality and that might be misleading to readers, depending on the text consulted and the problem the researcher wishes to solve."[3]

A second alternative is to remove enough variables from model 2.1 and — after the estimation — to transform the coefficients as suggested by Suits (1984). Thus, one applies OLS to estimate:

$$\begin{aligned} y = \tilde{\beta}_0 + \sum_{m=1}^{M} \beta_m x_m + \sum_{l=2}^{T} \tilde{\tau}_l d_l^{Time} \\ + \sum_{k=2}^{K} \tilde{\gamma}_k^* d_k^{Treat} + \sum_{k=1}^{K} \sum_{j=N_{k-1}+2}^{N_k} \tilde{\mu}_j d_j^{Match} + \varepsilon \end{aligned} \quad (2.5)$$

with the variables as in model 2.1. Model 2.5 is identified. It employs the restrictions:

$$\begin{aligned} \tilde{\tau}_1 &= 0, \\ \tilde{\gamma}_1^* &= 0 \qquad \text{and} \\ \tilde{\mu}_{N_{k-1}+1} &= 0, \forall k = 1, ..., K. \end{aligned}$$

I.e., the coefficients for period 1, treatment group 1 and the first matching group within each treatment group are set to zero and are left out in the formulation of the model. The tildes above the coefficients are used to distinguish them from those in model 2.1. The superscript star in $\tilde{\gamma}_k^*$ indicates that these coefficients are different from the $\tilde{\gamma}_k$'s, which result from imposing the restrictions 2.2 and 2.4. But, following Suits, the coefficients $\tilde{\gamma}_k$ and $\hat{\mu}_j$ which satisfy conditions 2.2 and 2.4 can be calculated as:[4]

$$\begin{aligned} \tilde{\gamma}_k &= \tilde{\gamma}_k^* + c_k \qquad \text{and} \\ \hat{\mu}_j &= \tilde{\mu}_j - c_k \qquad \forall j = N_{k-1}+1, ..., N_k \qquad \text{with} \\ c_k &= \frac{1}{N_k - N_{k-1}} \cdot \sum_{j=N_{k-1}+1}^{N_k} \tilde{\mu}_j \qquad \forall k = 1, ..., K\,. \end{aligned}$$

[3] Greene and Seaks (1991), p. 563.
[4] See Suits (1984) pp. 178-179.

These transformations are simple, but tedious. In addition to the coefficients their standard errors and test statistics need to be computed. Even more computing has to be done if the restrictions 2.2 and 2.3 shall be imposed. So, transforming the estimation results seems rather inconvenient.

2.4.2 Nested Effect Coding

There is a third alternative and it will be referred to as "nested effect coding", since it requires a special coding scheme for the dummy variables.[5] The following model is estimated using OLS:

$$y = \tilde{\beta}_0 + \sum_{m=1}^{M} \beta_m x_m + \sum_{l=2}^{T} \tilde{\tau}_l d_l^{Time} + \sum_{k=2}^{K} \tilde{\gamma}_k d_k^{Treat} + \sum_{k=1}^{K} \sum_{j=N_{k-1}+2}^{N_k} \hat{\mu}_j D_j^{Match} + \varepsilon \tag{2.6}$$

with

D_j^{Match} — dummy variable for matching group j,

$$D_{jit}^{Match} = \begin{cases} -1 \text{ if } i = N_{k-1}+1 \\ +1 \text{ if } i = j \\ \;\;0 \text{ otherwise} \end{cases}$$

$\forall j = N_{k-1}+2, ..., N_k$ and $\forall k = 1, ..., K$

and all other variables as in model 2.1. The value of D_j^{Match} is -1 for all observations on the first matching group within the treatment group to which j belongs. It is $+1$ for all observations on matching group j itself and it is zero for all other observations. Thus, the matching group dummies are effect–coded, instead being indicator–coded as in model 2.5. Furthermore, the coding is nested within the treatment groups.

Model 2.6 is identified and satisfies

$$\tilde{\tau}_1 = 0,$$

$$\tilde{\gamma}_1 = 0 \qquad \text{and}$$

$$\sum_{j=N_{k-1}+1}^{N_k} \hat{\mu}_j = 0, \forall k = 1, ..., K.$$

[5] See e.g. Kerlinger and Pedhazur (1973) or Bortz (1993). An application of this type of coding in the analysis of experimental panel data can be found in Königstein and Tietz (1998, see chapter 1).

To see the latter notice that the model does not include matching dummies for the first matching group within each treatment, i.e. for $i = N_{k-1} + 1$. Given such i and some period t, model 2.6 implies

$$y_{it} = \tilde{\beta}_0 + \sum_{m=1}^{M} \beta_m x_{mit} + \tilde{\tau}_t$$
$$+\tilde{\gamma}_k + \sum_{j=N_{k-1}+2}^{N_k} \left(-\hat{\mu}_j\right) + \varepsilon_{it} \; .$$

Thus the coefficient $\hat{\mu}_i$ for $i = N_{k-1} + 1$ is implicitly set to

$$\hat{\mu}_{N_{k-1}+1} = - \sum_{j=N_{k-1}+2}^{N_k} \hat{\mu}_j, \forall k = 1, ..., K$$

which implies

$$\sum_{j=N_{k-1}+1}^{N_k} \hat{\mu}_j = 0, \forall k = 1, ..., K$$

So, model 2.6 satisfies restrictions 2.2 and 2.4.

Accordingly, the OLS–estimation directly produces the coefficients of interest, their standard errors and test statistics. One might argue that the coefficients $\hat{\mu}_{N_{k-1}+1}$ remain to be computed. But firstly, the computational work is substantially reduced compared to model 2.5 and, secondly, in many cases this is not necessary at all. Namely, the coefficients $\hat{\mu}_j$ represent biases of the respective matching groups. They need to be estimated to account for repeated measurement, but in many cases their size and other statistics are not of interest.

2.5 Summary

One major reason for running experiments is to investigate how the behavior of subjects reacts to controlled changes in the environment. Accordingly, one of the main goals of the data analysis is to measure the influence of experimental treatments. Several problems that may arise herein were discussed, and it was suggested how to proceed with them. At a very general level these are the problems of strategic interaction and repeated measurement. On a more specific level, in estimating a fixed effects model, there are the problems of identification and interpretation as well as to choose the estimation procedure. I want to spent a few more words on the first two points.

Experimental data analysis has to regard the potential correlation in the data due to strategic interaction and repeated measurement. The two sources of correlation should be distinguished, since they may be dealt with differently. By choosing strategic interaction groups as units of analysis, which deliver time series of decisions, the data set is a cross–section of strategically independent time series. Appropriate econometric methods might take care of the correlation due to repeated measurement. One possibility is to estimate a fixed effects model, which was described here.

An alternative and extremely conservative procedure would be to aggregate all decisions of an interaction group without regard to the time they were taken. But this has some severe disadvantages. Firstly, the dynamic information is lost. Who has ever run a dynamic experiment knows that subjects' behavior might change during the experiment due to learning or experience. The dynamic pattern of behavior is therefore valuable information in its own right. Secondly, not controlling for the influence of time may lead to overestimation or underestimation of the influence of experimental treatments. Thirdly, using a sequence of decisions of each interaction group instead of a single data point is advantageous since it increases the number of observations and in most case the degrees of freedom in regression analysis. This is the case since the increase in observations is usually larger than the number of additional parameters that need to be estimated to account for the time effects and the matching group effects.

3. Convergence to Equitable Play in the Repeated Ultimatum Game with Advance Production

3.1 Introduction

Claims for distributive justice are common in political and economic disputes, and individual perceptions of what is a fair allocation are heterogeneous!

Both of these propositions hardly need proof. For example, Frey (1990, p. 144) argues that distributional rather than allocative aspects are in many cases the reason for governmental interventions into the economy. Also, within negotiations between unions and employers the unions claim on occasion that a proposal made by the employers, e.g. regarding a change in wages, is "socially unfair". Since comparing the well–being of workers and employers is a rather complex issue, and it is usually not hard to find a statistic that backs the claim.

One might think that verbal statements regarding distributive justice do not really influence the solution of the respective political or economic conflict. But, considering that they may influence the public's view of the respective problem, and furthermore that the public exerts social pressure upon the bargaining parties, it seems unlikely that equity arguments have no "real" effect. Empirical support for either view as well as for the heterogeneity of equity perceptions would be desirable, but is hard to find in field data.

However, such evidence might be found experimentally. Within the laboratory participants can be exposed to situations that allow for multiple equity standards. Different perceptions of equity may imply specific allocations, and clustering of data points at these allocations can be taken as evidence for the distribution of equity types within the subject population.

This paper reports such an experiment. It features the repeated ultimatum game with advance production; i.e., the social output that is to be distributed among the bargaining parties results from joint production. In real life it is natural that distribution conflicts can be described in terms of input–output–relationships. In contrast, there are not too many experiments, so far, on non–cooperative bargaining games with production.[1] Some examples are:

- Berg, Dickhaut and McCabe (1995): dictator game with one–sided production
- Hackett (1993): sequential bargaining with joint production
- Harrison and Hirshleifer (1989): public good allocation with joint production[2]
- Königstein and Tietz (1998):[3] one–shot ultimatum game with joint production
- Gantner, Güth and Königstein (1998):[4] one–shot ultimatum game and Nash demand game with joint production

Other studies on more complex games in which bargaining and production are part of a principal–agent relationship are Fehr, Gächter and Kirchsteiger (1997) and Güth, Klose, Königstein and Schwalbach (1998).

The study of Königstein and Tietz (1998) — henceforth referred to as KT — is a baseline experiment for the current study. It featured an ultimatum game with advance production in which subjects played six games, but faced each opponent only once. Here, in contrast, subjects play twelve games — respectively six games against the same oppenent. So, this study allows for more experience and repeated play against the same opponent. Its main hypothesis is that these factors induce convergence toward equitable play.

The results confirm this hypothesis. In contrast, behavior does not move toward but away from the game theoretic solution, which predicts a rather uneven and inefficient allocation. Furthermore, by applying hit rate analysis in the spirit of Selten and Krischker (1983), it is shown how to reduce the number of potential equity concepts based on their respective descriptive power. It turns out that the accessibility requirement for equity standards has to regard the limitations of the rationality of economic agents.

---

[1] E.g. most of the studies contained in the survey by Roth (1995) are games without production. An early investigation of the influence of hypothetical asymmetries in production upon bargaining decisions is Mikula (1972).

[2] This game involves no bargaining, but it is nevertheless closely related.

[3] See chapter 1.

[4] See chapter 4.

3.2 Repeated Bargaining with Advance Production

The repeated game consists of six repetitions of a base game featuring ultimatum bargaining with advance production.[5] A single play of the ultimatum game with advance production will be referred to as "period t base game" (BG), whereas the repeated games may be termed "RG1" and "RG2", repectively.

In period t, first, both players choose individual production quantities and thereby determine the joint return. Then player 1, the "demander", states how much of the return he asks for himself. Afterwards, player 2, the "responder", decides whether to reject or accept 1's demand. In case of acceptance, the return is split accordingly. Otherwise, the return is lost. In either case the players have to bear their respective production cost and the base game of period $t + 1$ follows.

In the experiment subjects faced the same opponent in all six BG's. They played two repeated games — thus, twelve BG's in total. Between the two repeated games opponents and the numeric specification of the game parameters (see below) changed.[6] The base game in period t is characterized as follows:

1. Players choose production quantities q_{1t} and q_{2t}, with

$$q_{it} \in \{0, 1, 2, ..., 30\}.$$

   Associated cost are

$$c_{it} = c(q_{it}).$$

   The return is

$$r_t(q_{1t}, q_{2t}) = \rho_1 q_{1t} + \rho_2 q_{2t}$$

   with ρ_i being the players' constant marginal productivities. The numeric specifications for the cost function and the return function are given below.
2. Player 1 is informed about r_t and chooses his demand d_{1t}, with

$$0 \leq d_{1t} \leq r_t .$$

3. Player 2 is informed about r_t and d_{1t} and decides about $\delta_{2t} \in \{0, 1\}$, with

[5] The instructions are contained in the appendix.

[6] RG1 exhibited the same parameter specification as in KT and thus allows a direct comparison of results.

Table 3.1. Marginal Productivities and Cost Functions for Both Repeated Games

RG1
$\rho_1 = 22 \qquad \rho_2 = 44$
$\tilde{c}(q_{it}) = \begin{cases} 0 & \text{, for } q_{it} = 0 \\ 15 + 12.5q_{it} - 1.73q_{it}^2 + 0.115q_{it}^3 & \text{, otherwise} \end{cases}$
RG2
$\rho_1 = 24 \qquad \rho_2 = 48$
$\tilde{c}(q_{it}) = \begin{cases} 0 & \text{, for } q_{it} = 0 \\ \frac{3}{4}\left(94 + 12.5q_{it} - 1.73q_{it}^2 + 0.115q_{it}^3\right) & \text{, otherwise} \end{cases}$

$$\delta_{2t} = \begin{cases} 0: & \text{rejection of } d_{1t} \\ 1: & \text{acceptance of } d_{1t} \; . \end{cases}$$

The base game payoffs are

$$\Pi_{1t} = \delta_{2t} \cdot d_{1t} - c_{1t}$$

and

$$\Pi_{2t} = \delta_{2t} \cdot (r_t - d_{1t}) - c_{2t} \; .$$

A player's repeated game payoff is the sum of his base game payoffs. The marginal productivities of both players and the cost schedule differed between RG1 and RG2. The cost for each production level $c(q_{it})$ were determined by evaluating a continuous cost function $\tilde{c}(q_{it})$ and rounding the values to integers. The parameters are displayed in table 3.1.[7] Thus, the return function and the cost function of RG2 are just positive affine transformations of the RG1 functions.[8] The parameters for RG1 are illustrated in figure 3.1. It displays marginal productivities and marginal cost (ordinate) for the range of production quantities between 0 and 30 (abszissa). For low (high) production levels marginal

[7] The cost schedules are contained in the appendix. For each repeated game, two versions of the cost tables were applied: a standard version and an extended version. The standard version listed the potential quantities and the associated total cost. In addition, the extended version listed marginal cost and average cost. Each version was applied to about half of the subjects. However, the different cost tables had no significant effect on the production decisions.

[8] Note, that multiplying return in RG1 by the factor $\frac{24}{22}$ yields return in RG2. Furthermore, adding a constant of 79 to the cost function of RG1 and multiplying this sum by $\frac{3}{4}$ yields the cost function of RG2.

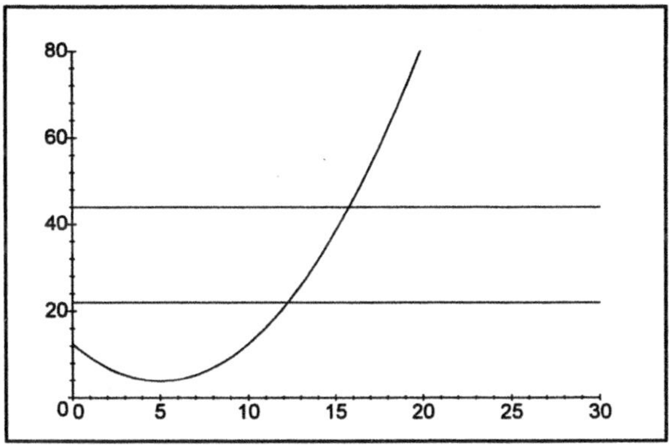

Fig. 3.1. Marginal Cost and Productivities (Ordinate) for Different Choices of Production Quantity (Abszissa) in RG1

productivities are higher (lower) than marginal cost. Thus, the efficient production levels are in the interior of the action space.

Return and cost are denoted in German Pfennig. Demands had to be stated in increments of one Pfennig. The numeric specification was designed to get a clear separation between the efficient production levels for both players and between both games.

3.3 Theoretical Analysis

Efficient production as well as behavior according to the unique strict subgame perfect equilibrium (SPE) implies for all periods t the results reported in table 3.2. The game theoretic solution results from solving the period 6 subgame and applying backward induction. In period 6 the remaining subgame is equivalent to a single play of the base game. It can easily be verified that the behavior as specified in table 3.2 is essentially the unique SPE–path of the period 6 subgame: Since player 2 gets 1 Pfennig he strictly prefers agreement to non–agreement and accordingly chooses zero–production. Player 1 receives almost all return and therefore chooses his efficient production level.

Since the decisions to be taken in period 5 cannot influence the period 6 equilibrium behavior, play is again as described in table 3.2. Accordingly, the game unravels from the last to the first period.

Equity theory can be applied to solve the distribution conflict characterized above. It predicts that a social output is allocated such that

Table 3.2. Some Theoretical Results for Both Games

	RG1		RG2	
Efficient Production:	q_{1t}	$= 12$	q_{1t}	$= 14$
	q_{2t}	$= 16$	q_{2t}	$= 18$
Efficient Surplus:	610		694	
SPE Behavior:	q^*_{1t}	$= 12$	q^*_{1t}	$= 14$
	q^*_{2t}	$= 0$	q^*_{2t}	$= 0$
	d^*_{1t}	$= r_t - 1$	d^*_{1t}	$= r_t - 1$
	δ^*_{2t}	$= 1$	δ^*_{2t}	$= 1$
SPE Profit:	Π^*_{1t}	$= 148$	Π^*_{1t}	$= 151$
	Π^*_{2t}	$= 1$	Π^*_{2t}	$= 1$

each player's received output share is equal to his input share, where a player's input is his contribution to social production.[9] However, it might be ambiguous how to measure inputs and outputs, and consequently players may disagree about "what is" a fair allocation, even if both want to implement equity. Regarding the base game, at first sight, there are ten reasonable candidates for an equity standard — i.e., a combination of a measure for inputs (input standard) and a measure for outputs (output standard). A player's input can be measured by his chosen quantity, his cost, his contribution to the joint return or his contribution to surplus. A further candidate for the input standard is the "equality standard", which proposes a "fifty–fifty" split of the output. For operationalizing the output standard, the return as well as the surplus are reasonable candidates.

Thus, in order to implement equity in the base game, the problem of selecting an equity standard arises. Since the game here is a repeated play of the base game, there is another source of ambiguity; namely, to implement equity in a dynamic sense. It could well be that in an experiment the subjects try to balance the repeated game payoffs, even though payoffs are not balanced in every single period.

[9] See Güth (1994), Homans (1961) and Selten (1978).

3.4 Experimental Procedures

The experiment comprised five sessions which were run during winter semester 1994/1995 at the University of Frankfurt/Main. In total, 56 subjects participated, mostly students of economics and business administration. They were not recruited from specific courses and were, in general, not familiar with game theory. In the laboratory the subjects were paired anonymously and edited their decisions into computers that were linked via a network. Subjects kept their respective role — demander or reponder in the ultimatum game — during the entire experiment and knew they were paired with different opponents in the two repeated games. After reading the instructions, subjects had to fill out a questionnaire regarding their planned decisions and expectations about opponent play and potential earnings.[10] It was used to check subjects understanding of the instructions. A post–experimental questionnaire — the 16–PA personality questionnaire of Brandstätter (1988) — was applied to collect some psychological characteristics.[11]

Mean earnings were about 24 DM per player (about 14.10 $ at the time of the experiment) for a 2 to 2.5 hours session including a 5 DM show–up fee. In case of efficient play subjects would have earned 44.12 DM. If subjects had played as prescribed by the game theoretic solution, player 1 would have earned 22.94 DM while player 2 would have earned 5.12 DM. Maximal earnings reached by a single subject were 27 DM. Four subjects incurred a total loss. Paying for total losses could not legally be enforced such that these subjects in fact only lost their show–up. But, since most subjects reached total wins, their losses occurring in some periods were booked against wins from other periods and the show–up fee, and thus most of the losses were indeed paid for by the participants.

3.5 Considerations of Experimental Design

The analysis will focus on the observable demand behavior. Since demands will be shown to depend on the production decisions, the latter will be also be part of the analysis.

Taking the KT study as a baseline, the experiment here features two treatments: repeated play against the same opponent — the "partner-

[10] The questionnaire is available from the author upon request.

[11] The post–experimental questionnaire is contained in the appendix.

ship" treatment — and experience. In RG1 everything is equal to KT except for the partnership treatment. Furthermore, RG2 is structurally identical[12] to RG1 except for subjects having gained experience. Thus, changes in behavior between KT and RG1 will be attributed to partnership and those between RG1 and RG2 to experience. As worked out above, theoretically the treatments should have no influence. But, assuming boundedly rational agents one may wonder about their consequences.

Intuitively, one might expect that the partnership treatment reduces the variance in the respective opponent's production choices, which a player observes in the course of the 6 BG's. Consequently, if the opponent's production choices vary less, a player might develop a clear bargaining pattern more easily. The same arguments hold for experience as well. Therefore, one may expect both treatments, partnership as well as experience, to reduce the noise in the data and bring out subjects' bargaining profiles more clearly.

On the other hand, one may hypothesize that equitable sharing vanishes. Subjects may need to learn about the strategic aspects of the game. As a consequence, one might observe less equitable play and more game theoretic decisions. Note that the former does not imply the latter. E.g. if a demander learns to exploit his first mover advantage to some degree his behavior moves away from equitable play, but is not as suggested by game theory.

Unless it comes to statistical testing, the analysis regards each base game as a single observation. In case subjects implement an equitable allocation of the repeated game payoffs, the base game split of the "pie" may seem unfair even though it is fair in the dynamic sense. The analysis does not account for this possibility. So, the predictive success of equity theory may be underestimated; the analysis is conservative.

3.6 Descriptive Statistics of Base Game Decisions

Table 3.3 reports some statistics on the base games for RG1, respectively RG2, and figures 3.2 to 3.5 display the distributions of production decisions. Of a total of 168 BG's[13] in RG1 (RG2) about 24% (17%) end in disagreement. The average surplus, i.e. return minus the sum of cost, is 3.93 DM in RG1 (5.93 DM in RG2). Thus, the efficiency

[12] Only the numerical specification differs.

[13] 28 pairs of players times 6 BG's.

Table 3.3. Statistics of the Base Games Within the Two Repeated Games

	RG1	RG2
Base Games	168	168
Agreements	127 (76%)	140 (83%)
Return per BG	8.37 DM	10.58 DM
Cost per BG		
Player 1	2.43 DM	1.90 DM
Player 2	2.01 DM	2.75 DM
Profit per BG		
Player 1	0.84 DM	2.37 DM
Player 2	1.11 DM	2.04 DM

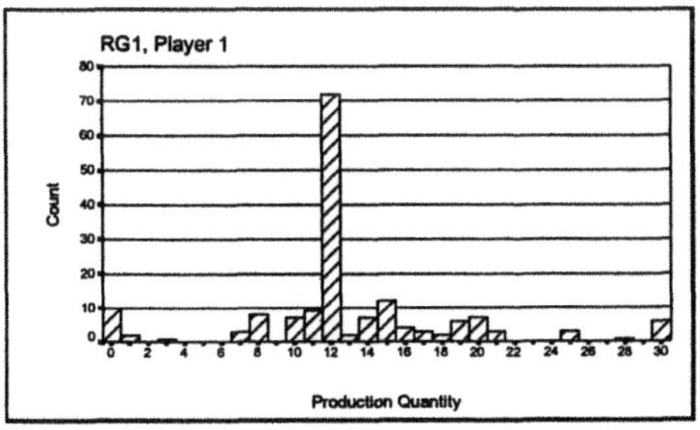

Fig. 3.2. Production Choices of Players 1 in RG1

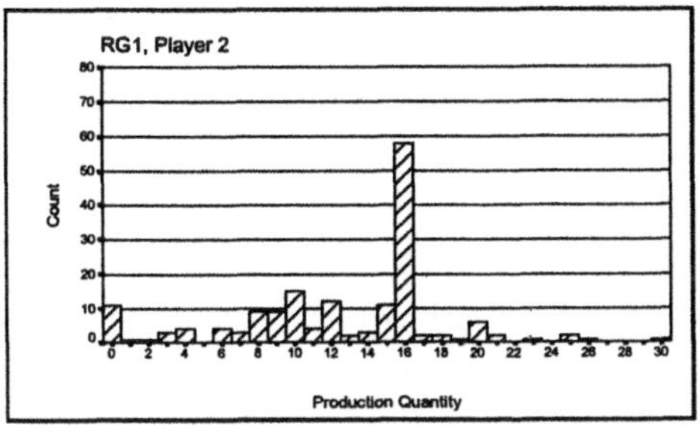

Fig. 3.3. Production Choices of Players 2 in RG1

rate of production is $\frac{393}{610} \approx 64\%$ in RG1 (about 85% in RG2). In RG1 player 1, especially, causes high inefficiency due to over–investment.

He invests 2.43 DM on average, while the efficient cost level is 1.15 DM. But, looking at the distribution of player 1's production choices one finds that nevertheless efficient production ($q_1 = 12$) is the modal decision and is chosen by a large number of subjects. Actually, about 43% of 1's production choices are efficient. So, the reported average

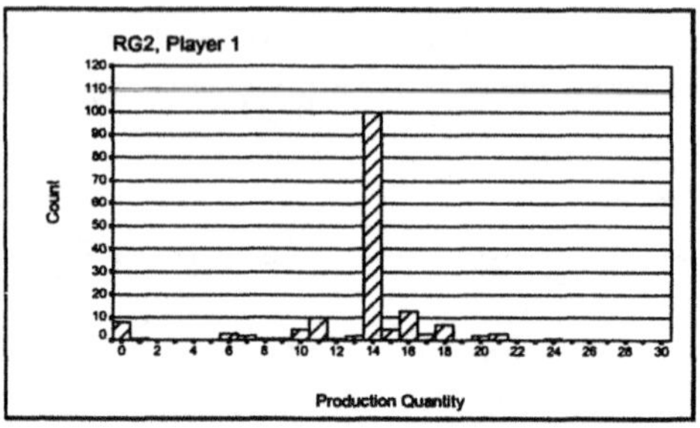

Fig. 3.4. Production Choices of Players 1 in RG2

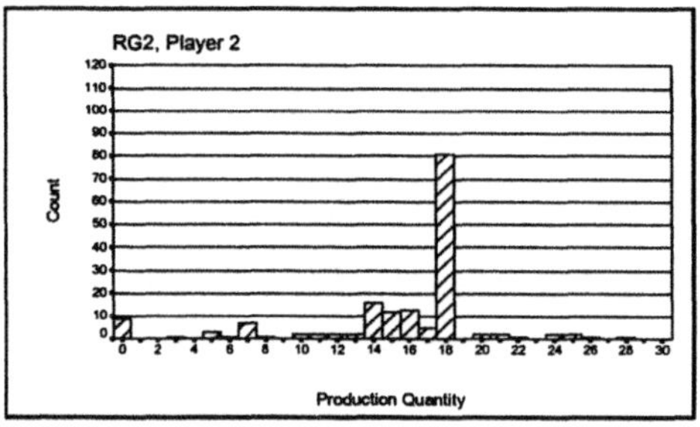

Fig. 3.5. Production Choices of Players 2 in RG2

efficiency is low mainly because there are some outliers at $q_1 = 30$. Since the cost function is convex, these observations strongly influence the average result.

The distributions for both players as well as both repeated games exhibit the respective efficient production level as modal decision. So, despite the difficult structure of the game a lot of subjects are able to determine and in fact choose the efficient production level. In RG2 about 60% of players 1 do so. Players 2 decide for efficient production in 36% (48%) of the cases in RG1 (RG2). Thus, experience leads to more productive efficiency.

As explained above, two different versions of cost tables were applied to about half of the subjects. Listing marginal cost and average cost in addition to total cost had no significant influence on first period production decisions.[14]

[14] By means of the Mann–Whitney U–Test and the Kolmogorov–Smirnov 2–Sample Test.

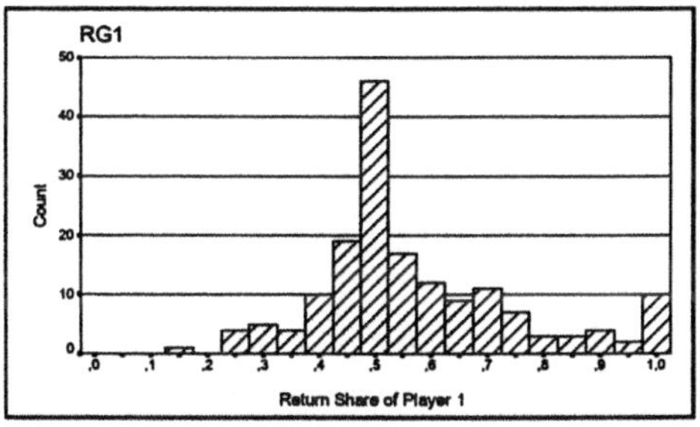

Fig. 3.6. Distribution of Return Shares Demanded by Players 1 in RG1

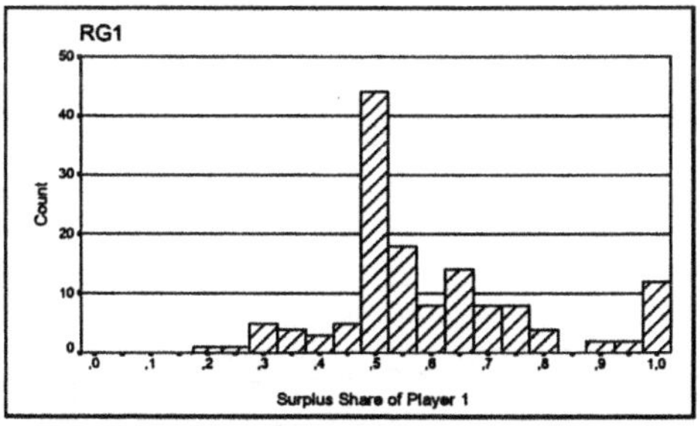

Fig. 3.7. Distribution of Surplus Shares Demanded by Players 1 in RG1

Figures 3.6 to 3.9 show the distributions of the demanded return shares as well as the demanded surplus shares.[15]

Accordingly, "fifty–fifty" is the modal demand in all cases. Experience seems to increase the importance of egalitarian considerations. Looking at the demanded surplus shares one finds that the modal class (where demanded shares are $0.475 \leq d_1 < 0.525$) exhibits about 31% of the cases in RG1 and about 53% in RG2.

Importantly, experience does not increase the number of greedy demands — e.g. demanded shares greater than 90%. Rather, the distribution moves away from extremly high demands. Not only the game theoretic solution receives little support, but also a "modified strategic

[15] Within the distributions of demanded surplus shares some observations are missing: The demanded surplus share was not computed if the contribution to surplus by one or both players was negative (24 cases in RG1, 3 cases in RG2). In case both players choose zero production (1 observation in RG1, 1 in RG2), neither the demanded surplus share nor the demanded return share is well defined. Furthermore, the figures do not display cases in which the demanded surplus share is smaller than zero (2 in RG1, 1 in RG2), and cases in which the demanded surplus share is greater than one (3 in RG1, 4 in RG2).

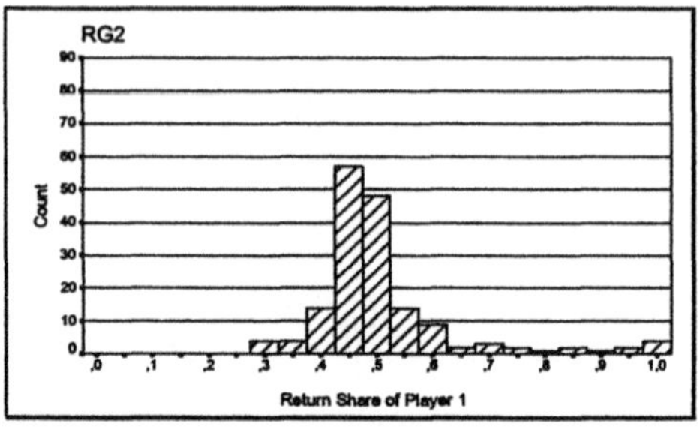

Fig. 3.8. Distribution of Return Shares Demanded by Players 1 in RG2

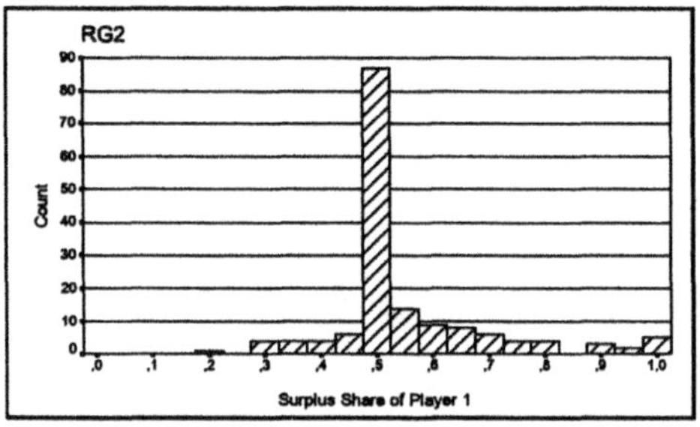

Fig. 3.9. Distribution of Surplus Shares Demanded by Players 1 in RG2

prediction"[16], which calls for demanding the whole surplus, thereby leaving player 2 nothing but his cost. According to table 3.3, experienced players 1 earn substantially higher profits on average than they would earn according to the game theoretic solution. Thus, non–equilibrium behavior does pay off.

A considerable amount of demanded shares are below 50%. This is different from what is usually observed in ultimatum games without production.[17] Such choices may be consistent with equity theory; namely, if player 1 bears a cost share lower than 50%. So, introducing "production" does make a difference in experimental studies on bargaining. The findings are summarized as follows:

Result 3.1 *Many subjects chose efficient production. Besides rejecting the game theoretic solution regarding player 2's behavior, this shows that most subjects clearly understood the game despite its complexity.*

[16] See KT.

[17] See e.g. Güth, Schmittberger and Schwarze (1982) and Forsythe, Horowitz, Savin and Sefton (1994).

Result 3.2 *Experience increases productive efficiency and furthermore increases overall efficiency due to a reduction in the rejection rate.*

Result 3.3 *The number of greedy demands is reduced via experience. So, behavior moves away from solutions suggested by strategic reasoning. In addition, experience seems to strengthen egalitarian considerations.*

3.7 Explanatory Power of Single Predictors

3.7.1 Hit Rate Analysis

Selten and Krischker (1983) proposed the following measure to determine the predictive success of area theories:

$$m = h - a$$

with h being the hite rate and a being the area of a theory that is investigated.[18] An area theory predicts a subset of the space of possible outcomes. The area is defined as the size of the predicted outcome space relative to the total outcome space. The hit rate is determined as the relative frequency of observations that belong to the predicted outcome space.

The measure can be used to compare the predictive success of various equity standards as well as strategic predictions. Originally, these concepts deliver point predictions for the demand of player 1. For empirical purposes they are in a sense "too precise". Experimental subjects might favor round numbers or might follow a behavioral pattern only approximately. Thus, a point predictor may hardly hit exactly, even if many subjects were guided by it. Consequently, one should count observations close to a predicted point as a hit of the respective theory.

But, then the question arises which prediction error to tolerate. It is not clear within which distance from the predicted point an observation should be considered a hit. The measure m suggests a solution to this problem, since it allows to consider different ranges and to compare the

[18] Selten (1989) discusses some mathematical properties of the measure and compares it to alternative measures.

results. Furthermore, instead of arbitrarily choosing a certain accuracy level one may choose the one that maximizes m.

The demanded return shares as well as the demanded surplus shares will be investigated by these means. A return share demanded by player 1 is a number within the unit intervall. By considering the unit intervall as partitioned into classes of width 1% one can measure the area of a theory as the percentage range predicted.[19] One could maximize m by considering all possible ranges, but instead it will be done only for all ranges up to 10%. Since the analysis suggests to combine point predictions with a tolerable prediction error, it seems adequate to impose an upper bound on the tolerable error.

3.7.2 Hit Rates and Success Measures in RG1

The results are displayed in tables 3.4 and 3.5. The columns represent the various input standards that might serve to operationalize equity. Table 3.4 considers return as output standard, while table 3.5 considers surplus. In addition the empirical success of the "modified strategic prediction" as well as the subgame perfect equilibrium is investigated. The upper rows of the tables present the hit rates (in percent) for several output ranges. E.g. if the range of 2% was considered, all observations of a demanded output share within a distance of ± 1 percentage points of the predicted output share were counted as hits. The rows below report the area a^* that maximizes the success measure, the corresponding maximal hit rate h^* as well as the maximal value of the success measure m^*; all three are expressed in percent.[20] As explained above, the maximization was done by increasing the range in steps of 1% up to a maximal range of 10% of the respective output, and by comparing the values of m.

Table 3.4 reports the "quantity standard" being the best predictor of the demanded return share in RG1 for all ranges. The hit rate is about 44% on the 10% range. Given player 1's share of quantity share

[19] For theories based on surplus the area is not equivalent to the range (see below).

[20] Note that for a player's demanded return share the total outcome space is the unit interval. Thus, for theories based on return the concept of area is identical to what was defined as the range. But, for theories based on surplus the two are different. In general, a player may demand a surplus share lower than 0 or greater than 1. Thus, a given surplus range, e.g. a 10% range, exhibits a relative size compared to the total outcome space that is smaller than 10%. Since the size of the surplus in RG1 is on average 61% of the return, the maximizing ranges were multiplied by 0.61 to yield the areas in table 3.5.

Table 3.4. Hit Rate Analysis for RG1 Data with Return as Output Standard

Range	Equality	Quantity	Cost	Contr. to Return	Contr. to Surplus	Subgame Perf. Eq'm
1%	10.8	11.4	7.2	4.2	1.4	4.2
2%	12.6	20.4	9.6	5.4	2.8	4.8
5%	27.5	27.5	16.2	7.8	3.5	6.0
10%	36.5	43.7	24.0	10.8	4.9	6.6
a^*	8.0	10.0	10.0	4.0	2.0	1.0
h^*	34.7	43.7	24.0	7.8	2.8	4.2
m^*	26.7	33.7	14.0	3.8	0.8	1.0

Table 3.5. Hit Rate Analysis for RG1 Data with Surplus as Output Standard

Range	Equality	Quantity	Cost	Contr. to Return	Contr. to Surplus	Modified Strat. Pred.
1%	21.5	5.6	4.9	0.0	2.1	5.6
2%	27.1	8.3	8.3	1.4	3.5	6.9
5%	30.6	15.3	12.5	2.8	4.9	8.3
10%	38.9	22.9	19.5	12.5	6.9	9.0
a^*	6.1	6.1	6.1	6.1	1.2	1.2
h^*	38.9	22.9	19.5	12.5	3.5	6.9
m^*	32.8	16.8	13.4	6.4	2.3	5.7

were 30% this number says that almost half of the player 1 population would demand a return share between 25% and 35%. The second best predictor of demanded return shares is the "equality standard".

In case the two players bear different cost, equal sharing of the return might nonetheless be viewed unfair. So, equal sharing of surplus may be considered instead. Table 3.5 shows that "equality standard" is an even better predictor of the demanded surplus share. It hits in 21.5% on the 1% range; these are exact hits, basically. For games without production other studies found substantially higher proportions of equal splits.[21] From a bounded rationality perspective this is not surprising. Splitting surplus equally requires much more cognitive effort when there is asymmetric production than with symmetric production (or none at all). In addition, one should consider heterogeneity of equity types. Without production equity standards based on return are equivalent to those based on surplus. Consequently, such games do not allow for separation between these standards.

[21] See e.g. Güth, Schmittberger and Schwarze (1982) as well as Forsythe, Horowitz, Savin and Sefton (1994). In games without production the "pie" to be distributed is necessarily a surplus.

For a comparison of theories based on return with theories based on surplus, one should look at m^*. Accordingly the success of the combination equality (input) and surplus (output) is measured as 32.8%, while that of the combination "equality–return" is 26.7%. "Equal surplus splitting" is theoretically more convincing than "equal return splitting"; and this is supported empirically. It should be noticed that the concepts are non–exclusive. If both players bear the same cost, both equity standards predict the same.[22]

Furthermore, m^* is maximal for the equity standard based on quantity (input) and return (output), which is a rather simple concept. It is certainly less relevant for equity in an economic sense than e.g. an equity standard that uses cost as input standard, but it is easy to implement. These findings support the importance of "accessibility" which according to Selten (1978) is a necessary property for an equity standard. They indicate that accessibility depends not only on objective information, but also on cognitive limitations of a player. Sophisticated equity standards based on the players' contributions to return or surplus as well as the game theoretic prediction and the modified strategic prediction predict rather poorly. The cost standard is substantially more successful than these, but substantially worse than the best predictors as well.

3.7.3 Hit Rates and Success Measures in RG2

Tables 3.6 and 3.7 display the hit rate analysis for RG2. Again the combinations "quantity–return" as well as "equality–surplus" are the two rules of implementing equity that dominate all others in terms of predictive success. It seems indeed remarkable that $\frac{2}{3}$ of the population demand a share within ±5 percentage points of the respective quantity share!

Comparing results between RG1 and RG2 one finds that, with experience, good predictors get even better, while bad predictors get even worse. Thus, experience leads behavior away from strategic reasoning or sophisticated equity considerations toward simple and egalitarian equity considerations.

[22] Furthermore, with cost as input standard the two output standards, return and surplus, predict the same in all cases; i.e. these concepts are equivalent. Also, the combinations "cost–return" and "cost–surplus" are equivalent; i.e., they imply the same demand d_1. Nevertheless, the results reported in tables 3.4 and 3.5 differ slightly, since the considered output ranges have different relative size (area) in the respective output space.

Table 3.6. Hit Rate Analysis for RG2 Data with Return as Output Standard

Range	Equality	Quantity	Cost	Contr. to Return	Contr. to Surplus	Subgame Perf. Eq'm
1%	19.8	26.3	12.6	1.2	1.2	1.8
2%	20.4	46.1	13.2	1.2	1.2	1.8
5%	28.7	56.9	19.8	1.8	1.8	1.4
10%	40.7	66.5	32.3	4.2	3.0	3.0
a^*	10.0	10.0	8.0	1.0	1.0	1.0
h^*	40.7	66.5	31.7	1.2	1.2	1.8
m^*	30.7	56.5	23.7	0.2	0.2	0.8

Table 3.7. Hit Rate Analysis for RG2 Data with Surplus as Output Standard

Range	Equality	Quantity	Cost	Contr. to Return	Contr. to Surplus	Modified Strat. Pred.
1%	38.2	13.9	12.1	3.0	1.2	1.8
2%	44.2	14.5	13.3	3.0	1.2	1.8
5%	52.7	19.4	15.2	4.2	1.2	3.0
10%	60.0	31.5	23.0	7.9	1.8	4.2
a^*	5.0	5.6	5.0	0.6	0.6	0.6
h^*	59.4	31.5	22.4	3.0	1.2	1.8
m^*	54.4	25.9	17.4	2.4	0.6	1.2

Table 3.8 presents another look at experience effects. In addition, it compares the results here with those of KT. It shows the maximal success measure m^* (in percent) for the data of KT, RG1 and RG2 for the four equity standards that were most successful in RG1. The measure is given for all base games, as well as separately for base games 1 and 2 versus 3 to 6.[23] The columns correspond to the equity standards "quantity–return", "equality–surplus", "equality–return" and "cost–return". Comparing the entries for the KT study with those for RG1 shows that introducing the repeated game treatment increased the predictive success of equity theory. Comparing RG1 and RG2 one finds a rather strong increase due to learning across different repeated games. And finally, by and large this dynamic effect works also within each repeated game confirming its robustness. The findings of this section shall be summarized as follows:

Result 3.4 *Among the candidates, the most successul predictor is equity theory based on quantity (input standard) and return (output stan-*

[23] The partitioning was chosen, since the regression analysis in KT showed that learning is strong in the beginning and stabilizes in periods 3 to 6.

Table 3.8. Success Measures of Good Predictors for Different Data Sets

	Quantity — Return	Equality — Surplus	Equality — Return	Cost — Return
KT				
All	28.1	24.4	19.3	15.2
1, 2	25.7	19.0	23.7	17.5
3 to 6	29.3	27.7	17.0	14.5
RG1				
All	33.7	32.8	26.7	14.0
1, 2	25.7	21.8	27.7	13.2
3 to 6	37.7	37.9	26.9	17.3
RG2				
All	56.5	54.4	30.7	23.7
1, 2	45.4	56.9	29.3	17.0
3 to 6	62.1	53.3	31.4	27.1

dard). About $\frac{2}{3}$ of the data for experienced subjects are consistent with this theory.

Result 3.5 *Equity based on equality (input standard) and surplus (output standard) is also quite successful. Since the predictions may overlap, there is some double–counting. Thus, the difference in the success of the two best predictors should be taken with care.*

Result 3.6 *Experience as well as repeated play against the same opponent improve the performance of these predictors and deteriorates that of sophisticated equity standards as well as that of strategic concepts.*

3.8 Joint Predictions and Convergence to Equitable Play

The hit rate analysis above does not control for double–counting. It determines the predictive power of each equity standard taken as a single predictor, but it does not tell much about the success of joint predictions. Therefore the following procedure is applied: each observation is exclusively assigned to its nearest neighbor among the four equity standards considered above, but only, if the distance is at most ± 2 percentage points of the respective return. Such observations are referred to as "equitable". All observations that are in further distance are classified as non–equitable. Thus, the joint prediction of the four

Table 3.9. Relative Frequencies of Equitable and Non–Equitable Demands

	KT	RG1	RG2
Equity Standard			
Quantity–Return	10.5	11.1	18.6
Equality–Surplus	16.8	20.1	33.5
Equality–Return	10.4	18.3	13.9
Cost–Return	8.9	8.2	6.0
Equitable	46.6	57.7	72.0
Non–equitable	53.4	42.3	28.0
$\sum$	100%	100%	100%
N	294	168	168

standards covers 16% (4 times 4%) of the possible outcome space. The relative frequencies of equitable versus non–equitable choices (in percent of all choices) as well as the distribution over the various equity standards[24] are shown in table 3.9. Accordingly, the percentage of equitable choices, which is 46.6% in KT, increases to 57.7% in RG1. In RG2 almost 3 out of 4 subjects choose an equitable demand. Looking at the proportions of the different equity standards we find equal surplus splitting as most common among subjects. It is substantially more often applied than the combination "quantity–return", which was most successful as a single predictor. So the predictive power of the latter combination as reported in table 3.8 may be overestimated to some extent. Nevertheless, the frequency of such choices is increasing by experience, which is not the case for the combinations "equality–return" and "cost–return" (RG1 versus RG2). It seems that there is a shift away from the latter equity standards towards the other two.

But, most importantly, the numbers indicate convergence to equitable play, which is a main issue of this study. Besides comparing frequencies, some formal statistical tests will be run to investigate the hypothesis:

Experience and repeated play against the same opponent ("partnership"), increase the frequency of equitable choices!

When it comes to testing one has to regard that the subjects interacted with one another and were observed repeatedly.[25] This problem was handled as follows:

[24] In some cases an observations was nearest neighbor to two or more standards. Such observations were proportionally assigned to the involved classes.

[25] For more on these problems within experimental data analysis see e.g. Davis and Holt (1993) and chapter 2.

Firstly, the data were aggregated by interaction groups. For the KT data this meant that all data of one session were aggregated to yield a single data point — the proportion of equitable demands — for each session (8 observations). Within the repeated games, RG1 and RG2, the subject pool of each session was subdivided into matching groups of 4 or 6 players (half of wich were players 1). So, each of the 9 matching groups delivered two related observations: the proportion of equitable demands in RG1 and in RG2.

Secondly, a Mann–Whitney U–Test was run on the interaction group data of KT versus RG1 (test of partnership effect), and a Wicoxon Matched–Pairs Signed–Ranks Test on the interaction group data of RG1 versus RG2 (test of experience effect).

Both tests reject the Null–Hypothesis on the 5%–significance level (one–tailed) and support the above hypothesis.

3.9 Summary

Selten (1978) argued that in a given game there may be multiple candidates to serve as equity standards. In addition he proposed two properties which a candidate for an equity standard should exhibit: "accessibility" and "relevance to the problem".

The non–cooperative game studied here, the repeated ultimatum game with advance production, does allow for competing equity standards. It turned out that many subjects implement an equal split of the surplus, but there is also as substantial proportion among the participants who follow a simple notion of equity; namely, combining "quantity" as input standard with "return" as output standard.

Compared to e.g. the players' cost, their chosen quantities are certainly less relevant to the problem. That nevertheless many subjects look at their quantity share, is probably due to its superior accessibility for boundedly rational agents. Note, that the subjects have all information at hand to determine the most sophisticated equity standards (among the alternatives considered here). Thus, accessibility of an equity standard does not solely depend on the informational structure of the game, but also on the cognitive limitations of human agents.

Regarding the investigated experimental treatments, the finding is:

Result 3.7 *Experience and partnership lead behavior toward equitable play.*

Of course, not all decisions are equitable and this result is not supposed to mean that non–equitable decisions will vanish completely, if games are long enough. But, the qualitative result is obvious.

From the viewpoint of an experimental economist, this study can in part be regarded a robustness check of a behavioral regularity, since some evidence for equitable play was found in other studies before. But, closing, I want to take a wider perspective and get back to some issues adressed in the introduction. Accordingly, regarding distributive justice, the laboratory evidence found here is:

> *Subjects do care for distributive justice. Their perceptions of equitable allocations are heterogenous and these perceptions do induce "real" economic effects. Furthermore, equity perceptions do not just depend on the objective informational structure of the respective conflict, but also on information processing by the agents.*

4. Equity Anchoring in Simple Bargaining Games with Production

4.1 Introduction

An interesting development in experimental studies of bargaining is that participants first must produce what they then can distribute via reaching an agreement. Examples for abstract games are, for instance, the studies by Berg, Dickhaut and McCabe (1995), Hackett (1993), Königstein (1998a)[1] as well as Königstein and Tietz (1998)[2]; situations where this is part of a principal agent problem are investigated by Fehr, Gächter and Kirchsteiger (1997) as well as Güth, Klose, Königstein and Schwalbach (1998).

Here, we will examine two types of simple bargaining games with advance production: the ultimatum game[3] and the "Nash–demand game". The latter is a novel experimental implementation of the demand game studied by Nash (1950).

Social production is asymmetric and additive, and it causes individual cost.[4] For such games, equity theory proposes that the social output should be distributed proportionally to the players' inputs. But, since it is not clear how to measure output — e.g. net of cost or not — and input — e.g. neglecting individual productivities or not —, subjects might disagree in what constitues a fair allocation. We show how the number of potential equity standards can be reduced.

Our main issue is to introduce a behavioral model we refer to as "equity anchoring" and to report some empirical support for it. It claims that subjects determine an allocation, which they consider as equitable, and which anchors their reasoning. They may overbid or undercut the anchor e.g. to adjust for strategic asymmetry.

[1] See chapter 3.

[2] See chapter 1.

[3] See Güth, Schmittberger and Schwarze (1982).

[4] A related study considering different social composition functions is Harrison and Hirshleifer (1989).

Uncovering such subtle aspects of bargaining behavior seems impossible within usual ultimatum game data. But, since we require two decisions from each subject, a demand and an acceptance bound, we allow to separate and therefore reveal both, an equity target and a tactical bargaining margin. We find 40% of the decision vectors consistent with equity anchoring. Another 30% of vectors are such that both components closely match equitable choices.

The paper is organized as follows: Section 2 explains the experimental games, the game theoretic solution and equity theory. Section 3 describes experimental procedures. Data analysis starts with a look at the raw data (section 4) and proceeds with an analysis of aggregate behavior (section 5). Section 6 introduces the concept of "equity anchoring" contains a more detailed analysis of individual decision behavior. Section 7 summarizes our findings.

4.2 Bargaining with Advance Production

4.2.1 Game Rules

The experimental games can be characterized as simple bargaining games with advance production.[5] That is, two parties, players X and Y, can jointly improve their situation by first producing a "pie" — an amount of money — and then distributing it. More formally, social production is as follows:

$$r = \rho_x q_x + \rho_y q_y$$

with

r	—	joint return
ρ_i	—	productivity of player i with $\rho_x = 3$ and $\rho_y = 6$
q_i	—	investment level (cost level) of player i with $q_i \in \{0, 1, ..., 50\}$.

Within the production phase of the game both players simultaneously choose their respective investments. The numbers q_i represent units

[5] The instructions which the subjects received are given in the appendix. Note, that some variable names are not identical between instructions and the text here.

of ECU ("experimental currency unit"), a fictitious currency with 10 ECU = 1 DM. Return r is determined according to the above function. Then, both players are informed about r — thereby about individual investments as well — and the bargaining phase of the game starts.

Here, each payer $i = x, y$ simultaneously chooses a demand d_i and an acceptance bound a_i such that

$$d_i \in \{0, 1, ..., r\}$$
$$a_i \in \{0, 1, ..., r\}$$
$$d_i \geq a_i \ .$$

Then the payoffs are determined according to the rules of the ultimatum game ("U–treatment") or the Nash–demand game ("N–treatment"). Subjects know whether they are playing under the U–treatment or the N–treatment. They are informed about their payoffs and the game ends.

The U-treatment means that payoffs are calculated according to the rules of the ultimatum game. An unbiased chance move determines whether player X (Y) becomes first mover — for now, we refer to the first mover as "player i" —, repectively second mover ("player j"). Then, the demand decision of the first mover d_i is matched with the acceptance bound of the second mover a_j. An agreement is reached if

$$r - d_i \geq a_j$$

— that is, if the first mover i offers the second mover j an amount at least as high as j's acceptance bound.[6] Otherwise, the game ends in disagreement.[7] The payoffs are

$$\Pi_i = \delta d_i - q_i \qquad \text{and} \qquad \Pi_j = \delta(r - d_i) - q_j$$

$$\text{with} \qquad \delta = \begin{cases} 1 & \text{in case of agreement} \\ 0 & \text{in case of disagreement} \end{cases}$$

since each player has to bear his cost in any case.

[6] This payoff rule is actually the reason for calling the a–decision the choice of an "acceptance bound".

[7] A similar game, namely an ultimatum game without production played "under the veil of ignorance" and with players submitting strategies for both roles was already investigated by Güth, Schmittberger and Schwarze (1982) and Harrison and McCabe (1992).

The N–treatment means that payoffs are calculated according to the rules of what we refer to as the Nash–demand game. In fact, it is not exactly the game studied by Nash (1950), since we collect two decisions, d_i and a_i. The way in which agreements are reached is different, but nevertheless similar. Namely, we distinguish between a d–agreement — meaning that demands are compatible — and an a–agreement — implying compatible acceptance bounds. Formally, the conditions are:

$$\begin{array}{ll} d\text{–agreement:} & d_x + d_y \leq r \\ a\text{–agreement:} & d_x + d_y > r \wedge a_x + a_y \leq r \\ \text{disagreement:} & \text{otherwise.} \end{array}$$

The payoffs are

$$\Pi_i = \begin{cases} d_i - q_i & \text{for } d\text{–agreement} \\ a_i - q_i & \text{for } a\text{–agreement} \\ -q_i & \text{if disagreement} \end{cases}$$
$$\text{for} \quad i = x, y$$

Again, each player has to bear his cost in any case. Furthermore, in case of agreements any residual money not claimed by the players — i.e., $r - d_x - d_y$ (d–agreement) or $r - a_x - a_y$ (a–agreement) — is lost.

4.2.2 Game Theoretic Solution

The game theoretic solution to the bargaining phase of the ultimatum game will be derived by applying the perfect equilibrium concept[8]. Suppose in case of indifference the second mover prefers an agreement. Then, the unique solution is to choose

$$d_i = r \qquad \text{and} \qquad a_i = 0 \qquad \text{for } i = x, y.$$

Within the Nash–demand game, all strategy combinations

$$((d_x, a_x), (d_y, a_y)) \qquad \text{with } a_x + a_y = r$$

constitute an agreement equilibrium. There exist one disagreement equilibrium, given by $d_i = a_i = r$ for $i = x, y$. We select a unique

[8] See Selten (1975).

agreement equilibrium as the solution of the game by employing the co-operative bargaining solution of Nash (1950). It is the agreement equilibrium $\left((d_x, a_x^*), (d_y, a_y^*)\right)$ that maximizes the Nash product. Thus, the solution requires

$$a_x^* \cdot a_y^* \geq a_x \cdot a_y$$

$$\text{for all } a_x, a_y \text{ such that } a_x + a_y = r.$$

Since all decisions had to be integer numbers, the solution is $a_x^* = a_y^* = \frac{r}{2}$ if r is even. If r is odd, a_x^* and a_y^* can differ at most by 1.

Regarding production we consider the Nash– demand game first. The marginal return received by player i from investing one monetary unit is $\frac{1}{2}\rho_i$ which is constant and greater than marginal cost. Thus, $q_x^* = q_y^* = 50$ are the equilibrium production choices. In the ultimatum game, optimal production requires $q_i = 50$ for the first mover and $q_j = 0$ for the second mover. But, since the roles are not known in advance the production choice should maximize i's expected utility. If a player is not too risk averse, he should choose 50; specifically, a risk neutral player should do so.

4.2.3 Equitable Allocations

Equity theory (see, for instance, Homans, 1961) proposes that a social output should be split proportionally to the players inputs. More specifically, a player's output share should equal his input share. In practice, it might be ambiguous what to consider as social output and as a players' inputs. Thus, determining an equitable allocation requires to operationalize outputs and inputs. It will be referred to as choosing an output standard and an input standard or, generelly, as choice of an equity standard, that is, a combination of input– and output standard. In the games studied here one might e.g. consider return r — i.e., gross output — or, alternatively, surplus $s = r - q_x - q_y$ — i.e., net output. A player's input could be measured by his cost, his contribution to the joint return or his contribution to the joint surplus. In addition, to these there is the so called "personal contribution standard",[9] which measures total input by the number of persons involved in the game and proposes to give each person an equal share of the output. Thus, it allows to model egalitarian considerations within the more general framework of equity theory. It will simply be referred to as "equality

[9] See Güth (1988).

Table 4.1. Input Standards Based on Player i's Cost, his Contribution to Return or Surplus as well as the Equality Standard

Input Standard		Input Share
Cost:	q_i	$\gamma^q \equiv \frac{q_i}{q_i+q_j}$
Contr. to Return:	$\rho_i q_i$	$\gamma^r \equiv \frac{\rho_i q_i}{\rho_i q_i+\rho_j q_j}$
Contr. to Surplus:	$\rho_i q_i - q_i$	$\gamma^s \equiv \frac{(\rho_i-1)q_i}{(\rho_i-1)q_i+(\rho_j-1)q_j}$
Equality:	1	$\gamma^e \equiv \frac{1}{2}$

standard". The various candidates for the input standard and the corresponding input shares are listed in table 4.1. A vector $\left(\gamma^k, l\right)$, with $k \in \{q, r, s, e\}$ and $l \in \{r, s\}$ as defined above, denotes an equity standard comprising input standard γ^k and output standard l. Ambiguity with respect to the equity standard implies, that even if both players want to implement an equitable allocation, they nevertheless might disagree upon "what is equitable". If a subject applies a certain equity standard (γ^k, l) we may refer to this as the subject's equity type. We will determine the solution for homogeneous equity types. According to equity theory the solution to the bargaining phase of the game is

$$d_i = a_i = \begin{cases} \gamma^k \cdot r & \text{if } i = x, y \text{ are of type } \left(\gamma^k, r\right) \\ q_i + \gamma^k \cdot s & \text{if } i = x, y \text{ are of type } \left(\gamma^k, s\right) \end{cases}$$

and is the same for the ultimatum game and the Nash–demand game. Thus, for homogeneous equity types, the game always ends in agreement and payoffs are

$$\Pi_i = d_i - q_i \qquad \text{for } i = x, y.$$

The marginal return received by player i is non–linear in q_i — via γ^k for $k = q, r, s$. But, notice that equity based on surplus as output standard implies that each player gets at least his cost. Furthermore, γ^k, r and s are increasing in q_i such that all types $\left(\gamma^k, s\right)$ should choose efficient production.

The same holds for type (γ^e, r), since it prescribes the solution of the Nash–demand game. Finally, the payoffs of types (γ^r, r), (γ^s, r) and (γ^q, r), are

$$\Pi_i\left(\gamma^r, r\right) = \rho_i q_i - q_i$$
$$\Pi_i\left(\gamma^s, r\right) = \left(\rho_i - 1\right) q_i \cdot \frac{r}{s} - q_i$$

$$\text{and} \qquad \Pi_i(\gamma^q, r) = q_i \cdot \frac{r}{q_i + q_j} - q_i$$

which are always positive, unless $q_x = q_y = 0$, and increasing in q_i — notice that $\rho_i > 2$ and that $\frac{r}{s}$ and $\frac{r}{q_i+q_j}$ are at least 1. Following these arguments, $q_i^* = 50$ is optimal for all equity types.

4.2.4 Some Properties of the Games and the Experimental Design

One should not expect that subjects implement one of these solutions precisely, but rather anticipate some dispersion in behavior. Thus, we want to consider some qualitative properties of the different predictions. While the Nash–demand game is strategically symmetric and has a symmetric solution, the ultimatum game is asymmetric. So, even if subjects do not implement the game theoretic solution exactly, one should expect less extreme bargaining decisions in the Nash–demand game.

Since the ultimatum game is played under the veil of ignorance with respect to the bargaining role, it may appear less asymmetric than an ultimatum game with known role assignments. But, it could well be that subjects exploit their first mover advantage even stronger than usually[10], since even the asymmetric equilibrium decisions implement equal (ex ante) expected payoffs.

The production technology is linear. Thus it is simpler than in the experiments of Königstein (1998a) as well as Königstein and Tietz (1998).[11] Since even in those studies with non–linear cost a substantial amount of equitable play was observed, one might expect the data here to even closer fit equity theory.

According to the theoretical analysis there are eight candidates to serve as equity standards — four input standards combined with two output standards. One can easily verify that the standard (γ^q, r) is equivalent to (γ^q, s) and also that (γ^r, r) is equivalent to (γ^s, s). Thus we need to consider only six distinct candidates. Furthermore, for given production choices the various input standards predict different allocations, in general, but they may be equivalent for certain choices: e.g. $q_x = q_y$ implies $\gamma^q = \gamma^e = 0.5$.

[10] That subjects do exploit their first mover advantage to some extent, is a stylized fact of ultimatum games — see e.g. Roth (1995).

[11] See chapters 1 and 3.

Table 4.2. Payoffs of Both Players Implied by Different Theoretical Concepts

	Player X	Player Y
Game Theory		
Ultimatum Game	400	−50
Nash–demand game	175	175
Equity Theory		
(γ^q, r)	175	175
(γ^e, r)	175	175
(γ^r, r)	100	250
(γ^s, r)	79	271
(γ^e, s)	175	175
(γ^r, s)	117	233

The empirical analysis below will focus on the bargaining decisions. Table 4.2 summarizes the theoretical predictions for the players' payoffs assuming efficient production; i.e., $q_x = q_y = 50$ and thus $r = 450$, $s = 350$.

4.3 Experimental Procedures

The experiment comprised two sessions run in summer semester 1996 with a total of 38 participants (undergraduate students of economics and management science at the Humboldt–University at Berlin). Each subject played two games, once the ultimatum–treatment and once the Nash–treatment. We employed two sequencing conditions: 22 subjects played "Ultimatum—Nash"; 16 subjects played "Nash—Ultimatum".[12] Subjects were randomly matched pairwise and kept their type, X respectively Y, in both games. Play was anonymous. Decisions were entered into decision forms, and the forms were collected by the experimenter. Matching of decisions and calculations were processed through a computer, and the forms were redistributed to give the feedback as prescribed by the game rules. Each session ran as follows:

1. Distribution of the two–page instuctions (see appendix). The instructions for the U–game and the N–game differ only in the second page.
2. After reading the instructions the participants could privately ask questions.

[12] The instructions in the appendix correspond to the sequencing "Ultimatum—Nash". Those for "Nash—Ultimatum" are analogous.

3. Distribution of the decision forms (see appendix), which were also used later to inform the players about r.
4. The forms were collected after the production decisions (first field of form) and distributed again after determining r fo r each pair of players and filling in r (second field of form).
5. After the subjects had filled in their bargaining decisions (third field of form) the forms were collected again. The bargaining result was determined and then the forms were distributed again to inform the subjects (fourth field of form). In the ultimatum game they were told their final payment and whether they were selected first or second mover. In the Nash–demand game, besides their final payment, they were told whether a d– or an a–agreement had been reached.
6. Distribution of a control questionaire (see appendix) in order to check whether the instructions were well understood.
7. Second game with changed pairings subject to the respective other bargaining rule (equivalent to stages 1 to 6).
8. Distribution of the 16–PA personality questionaire — see Brandstätter (1988) —, which is designed to determine some psychological characteristics.[13] In addition we asked for age group, gender, field of study, number of semesters studied and received degrees.
9. Participants were paid privately. Within the session they received a code number which they had to show when cashing earnings at the secretary's office.

Average earnings were about 45 DM — about 29 $ at the time of the experiment — including a show–up fee of 20 DM. Sessions took about 75 minutes such that mean earnings were about 1.5 times the rate of a student job. The lowest payoff was 26 DM, the highest was 66 DM.

4.4 Empirical Distributions of Production and Bargaining Decisions

Figures 4.1 and 4.2 show the distributions of production choices for games 1 and 2.[14] We see a substantial proportion of efficient decisions $q_i = 50$. We found no significant differences between player types, X

[13] For an application of this questionaire in the analysis of ultimatum games see Brandstätter and Königstein (1998).

[14] All data are available from the authors upon request.

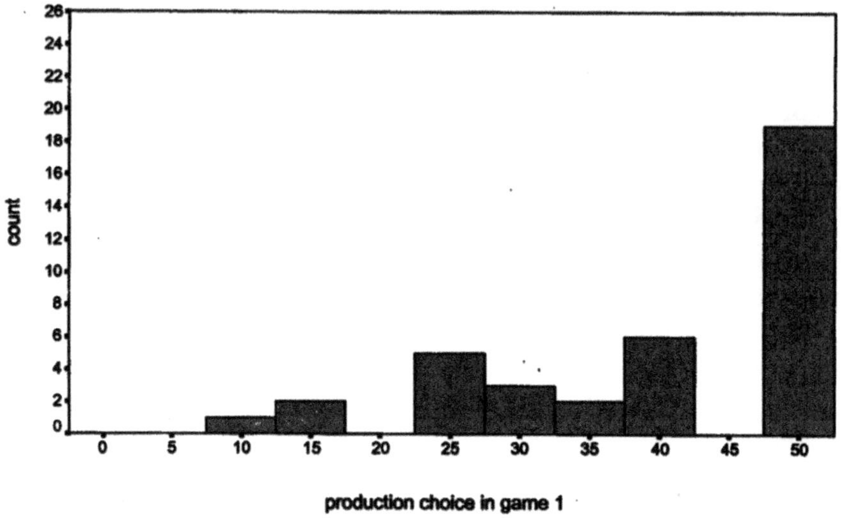

Fig. 4.1. Distribution of First Game Production Decisions

and Y, as well as between game types, U–game versus N–game.[15] In the first game $q_i = 39.8$ on average, while the mean is $q_i = 44.7$ in game 2. The increase in productive efficiency is statistically significant according to the Sign–Test ($\alpha < .001$, two–tailed).[16]

Figure 4.3 is a scatterplot of the combinations of bargaining decisions, d_i and a_i, relative to the available return; i.e. we calculated the demanded return shares $\frac{d_i}{r}$ as well as the acceptable return shares $\frac{a_i}{r}$. That all points cluster in the lower right part of the figure is implied by the restriction $a_i \leq d_i$. Most entries represent single data points, but the entry at $(d_i, a_i) = (0.5, 0.5)$ represents seven cases.

According to the game theoretic solution we should only observe points $(1, 0)$ — in the U–game — and $(d_i, 0.5)$ — in the N–game. Contrary to that, we observe no extreme demands and considerable

[15] We applied the Mann–Whitney U–Test as well as the Kolmogorov–Smirnov 2–Sample Test upon first game decisions.

[16] In statistical testing we had to consider that individual decisions are non–independent due to strategic interaction (see chapter 2). The Sign–Test is designed to handle repeated measurement, but not the potential correlation of game 2 decisions arising from player i having interacted with player j in game 1. Therefore we calculated the mean of q_i–decisons by each first game pair of players — i.e. we got a mean of q_i for each pair and each game. By choosing the player pairs as units of analysis for the Sign–Test, the strategic interaction problem is avoided.

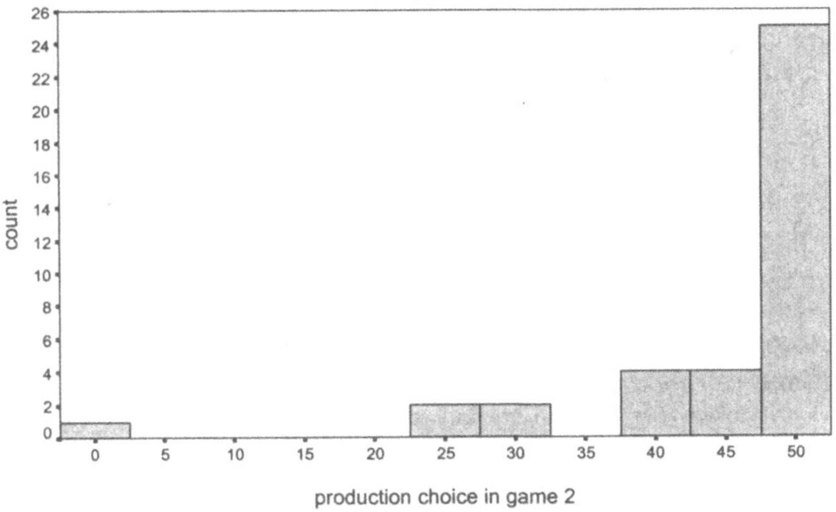

Fig. 4.2. Distribution of Second Game Production Decisions

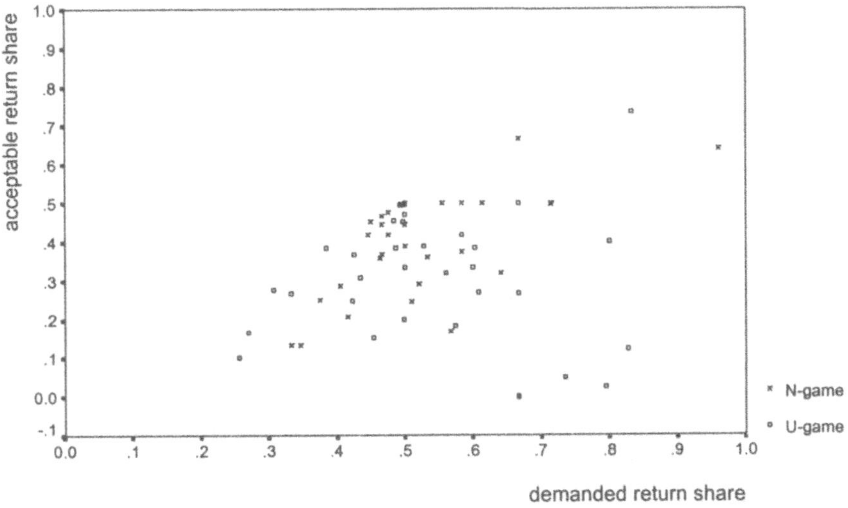

Fig. 4.3. Distribution of Bargaining Decisions for Both Types of Games

dispersion. The maximal demanded return share is 0.83. There are 6 cases (16%) with acceptable return shares below 0.10 — all of these are U–game data. In the Nash–demand game 8 cases (21%) of decisions $(d_i, 0.5)$ occurred, giving some support for the game theoretic solution. But, one should keep in mind that such decisions may rely on equity considerations if "equality" is the input standard and "return" the output standard. Regarding the relevance of other equity standards the figure is non–informative. To find out about this requires a detailed analysis (see below). The findings are summarized as:

Result 4.1 *Subjects choose efficient production choices in many cases. Efficiency increases with experience.*

Result 4.2 *By far the most bargaining decisons are inconsistent with the game theoretic solution — if taken strictly.*

4.5 Explaining Aggregate Behavior by a Single Equity Standard

As a first step to investigate whether subjects' decisions follow equity considerations we apply regression analysis. Both decisions, d_i and a_i, may be chosen equitably. So, we estimated the following model:

$$\theta_i = \beta_0 + \beta_1 \gamma_i^r + \varepsilon_i \qquad \text{for } i = 1, ..., N$$

with either using player i's demanded return share as dependent variable θ_i or his acceptable return share.[17] The model claims that θ_i, i.e., the demanded, respectively the acceptable return share, of player i, is a function of i's contributed return share γ_i^r and some random component ε_i.

Game theory produces a precise prediction for the model. In the ultimatum game, $(\beta_0 = 1, \beta_1 = 0)$ should hold, and in the Nash–demand game the coefficients should be $(\beta_0 = 0.5, \beta_1 = 0)$.

Equity theory predicts $(\beta_0 = 0, \beta_1 = 1)$ if all subjects exhibit type (γ^r, r), and $(\beta_0 = 0.5, \beta_1 = 0)$ if types are (γ^e, r). If the subject population consist of a mixture of the two types one should observe $(\beta_0 \in [0, 0.5], \beta_1 \in [0, 1])$ — more precisely, the coefficients for the mixed population should be a linear combination of the coefficients

[17] I.e., we ran separate regression on the two decision variables.

for the two types. For other equity standards the prediction is not that clear, but since return and surplus as well as γ^r, γ^q and γ^s are to some extent positively correlated one should definitely expect a positive coefficient β_1.

Even though we anticipate heterogenous equity types, it is interesting to investigate whether a single equity standard has an important influence on the decisions and can explain a substantial proportion of the variance in the data.[18] Table 4.3 reports the results for games 1 and 2. It displays the estimated coefficients and their standard errors (in parentheses). One (two, three) asterisks denote one–tailed significance at the 10%–level (5%, 1%).

In principle, the players' productivities could influence the decisions otherwise than just through γ_i^r. In addition, one might expect that the type of game that is played — ultimatum game versus Nash–demand game — is an important control variable. Therefore we first ran regressions in which the intercept or slope coefficients were estimated separately for each combination of player and game type and tested whether they were significantly different from one another. In general, this was not the case therefore we report regressions were only single coefficients are estimated. But, separate coefficients had to be estimated within regressions of the a–decisions, since one group of subjects — players Y playing the U–Game first and the N–Game second (coefficient β_1^+) — differed significantly from the rest of subjects (coefficient β_1).

[18] It should be noted that inclusion of more than one equity standard would be problematic due to multicollinearity. Furthermore, the dependent variable θ_i is bounded below and above such that the usual assumptions for statistical testing are not satisfied. But, since all d–decisions and almost all a–decisions are in the interior of the unit intervall with substantial distance to the boundaries we view the linear model as a good approximation. Another statistical problem might be caused by each of the two players of one pair observing the same value γ_i^r which could, in principle, result in correlated errors (see chapter 2). To avoid this one could analyze mean decisions by each pair. However, since the decisions of one group of players Y differed from the rest (see below), we think that the way we analyze the data here preserves more information. Besides, one can be conservative on testing and determine the significance levels using the degrees of freedom for paired data. Note, that pairing the data reduces the variance such that the T–values would increase. But, even by looking at the displayed T–values, one finds that basically nothing would change.

Regarding the tests for regressions on game 2 data, one should be more careful. The history (game 1) could have an influence. But it was left out in the model in order to better compare the result to the regressions on game 1 data.

Table 4.3. Regression Analyses for Both Decisions and Both Plays

	First Game		Second Game	
	$\frac{d_i}{r_i}$	$\frac{a_i}{r_i}$	$\frac{d_i}{r_i}$	$\frac{a_i}{r_i}$
Coefficients				
β_0	.335***	.164**	.389***	.265***
	(.062)	(.073)	(.046)	(.056)
β_1	.419***	.437***	.262***	.347***
	(.116)	(.153)	(.087)	(.114)
β_1^+		.043		.148*
		(.133)		(.105)
Model				
F	13.01***	9.14***	9.17***	5.31***
R^2_{adj}	.245	.305	.181	.189

In all regressions reported in table 4.3, we find that the model fits the data well, explaining about 20% to 30% of the variance. The intercept terms β_0 as well as the main slope coefficients β_1 are economically substantial and highly significant — i.e., different from zero. The intercept terms are below and significantly different from 0.5 as well. Thus, the model clearly rejects game theoretic behavior. It is inconsistent with the solution of the Nash–demand game and in even stronger contrast to the solution of the ultimatum game. It appears more consistent with equity theory. But, the contribution–to–return standard alone is not enough: the proposition ($\beta_0 = 0, \beta_1 = 1$) is rejected.

Result 4.3 *On the aggregate level, concern for equity seems to be a driving force of bargaining decisions. Even a model based on a single equity standard* (γ^r, r) *explains a considerable share of the variance.*

Result 4.4 *That* (γ^r, r) *is an economically substantial and statistically significant predictor, implies that asymmetries in players' productivities are taken into account by the bargaining parties.*

4.6 Equity Anchoring and the Tactical Bargaining Margin

4.6.1 Theoretical Considerations

Given an equity standard, equity theory proposes a precise distribution of the output. Since participants in an experiment favor round numbers and may determine equitable splits only approximately due to internal (calculation) cost, small deviations from equitable allocations should not be regarded as counterevidence for equity theory. But, there are substantial deviations as well.

As equity theory was laid out here, it neglects other potentially relevant aspects of the conflict, like e.g. strategic asymmetry or risk attitudes of the players. From a theoretical point of view, it is certainly desirable to contrast different concepts as much as possible, but in practice subjects' behavior might be driven by several forces instead of a single one. So, intuitively, one may wish to explain observable behavior by relying on several concepts. From an empirical viewpoint, however, this is useful only if it generates testable predictions.

We propose "equity anchoring" as a mixed behavioral model, that does generate testable predictions. Accordingly, equity theory suggests an allocation which anchors the decision of an individual, but the individual may overbid or undercut the anchor. Undercutting might reflect discounting for risk or inferior power — in case of a strategic disadvantage. Overbidding may result from risk loving or from a strategic advantage. Overbidding or undercutting of the equity anchor can be seen a "tactical bargaining margin". The examples we gave are potential causes for a tactical bargaining margin. However, our experiment is not designed to find out about these causes, but to find out in the first place, whether equity anchoring exists at all.

Assume for a moment that a subject decides either about d_i or a_i, but not about both at the same time — this is the usual case. Here, a deviation from an equitable decision might be attributed to equity anchoring, but it could mean as well, that the subject considered the game theoretic solution and implemented a risk discount. It is not possible to separate between these alternative explanations.

Now, consider our experiment, where each subject takes both decisions at the same time. It may be that both decisions aim to implement equity, which we will regard as "purely equitable" behavior. But, it might as well be that only one of the two decisions is equitable while the other overshoots or undercuts an equitable allocation. Such decision vectors will be taken as evidence for equity anchoring. Furthermore, if one observes vectors where at least one component is consistent with the game theoretic solution, this counts as evidence for strategic behavior. Of course, some observations can not be classified according to these categories.

4.6.2 Hit Rate Analysis

Königstein (1998a) and Königstein and Tietz (1998) applied a hit rate analysis to find out which equity standards are good predictors.[19] We proceed in the same way to reduce the number of equity standards to be considered later on. Some of the six candidates can be excluded for theoretical reasons (see below). We will see that theoretically unreasonable candidates also receive little empirical support.

Hit rate analysis proceeds as follows: An equity standard delivers a point prediction for the output allocation. Considering observations in a small neighborhood as a hit, one can count the total number of hits and calculate the hit rate, i.e the relative frequency of hits.

Of course, the hit rate increases with the size of the neighborhood. Therefore, a meaningful measure of predictive success m must take into account the hit rate h and the prediction area a, with the area being the relative size of the predicted outcome space.[20] Following Selten and Krischker (1983) we will measure predictive success as

$$m = h - a.$$

We will calculate m for different areas and then choose the maximal value m^* to represent the predictive success.

Specifically, we determine the predicted return share (surplus share) implied by a given equity standard. Then we increase the area from 1% up to 10% of the return (surplus) in steps of 1% of return (surplus). So we get the predictive success of the considered equity standards for different levels of accuracy. An accuracy level of 10% means that an observed output share within a distance of ±5 percentage points of the predicted output share is counted as a hit. Among the success measures for ten different accuracy levels we choose the maximal values. Table 4.4 reports the success (in percent) of various equity standard as well as the game theoretic solution in predicting demands and acceptance bounds in both types of games (and in parentheses the area that maximizes m). The success of (γ^q, r), i.e. equity based on cost (as input standard) and return (as output standard) is, for instance, 29.5 percent. This means that 39.5% of all demands in the ultimatum game cluster within ±5 percentage points (area: 10%) of the return allocation predicted by this equity standard — since: $hit\ rate = m^* + area$.

[19] See chapters 1 and 3.

[20] For a formal definition of the area and a comparision of alternative success measures see Selten (1989).

Table 4.4. Maximal Values for Predictive Success of Various Equity Standards and the Game Theoretic Prediction

	Ultimatum Game		Nash–demand game	
	d_i^u	a_i^u	d_i^n	a_i^n
Equity Standard				
(γ^q, r)	29.5	10.2	38.4	6.9
(γ^e, r)	28.2	13.8	45.3	27.2
(γ^r, r)	15.4	14.8	22.7	15.4
(γ^s, r)	3.2	8.8	11.1	8.4
(γ^e, s)	27.3	10.1	42.3	24.7
(γ^r, s)	8.1	-0.1	16.0	13.4
Game Theory	-1.0	6.9		27.2

Alternatively, it can mean that 31.5% of demands cluster within ±1 percentage points (area: 2%) of the return allocation.

Comparing the entries in the first column one finds that (γ^q, r), (γ^e, r) and (γ^e, s) are substantially better predictors than the rest. This result holds qualitatively for the Nash–demand game as well. The standards (γ^e, r), (γ^e, s) and (γ^q, r) predict substantially better than the rest, even though their rank order has changed.

Comparing the entries between game types we see that equity theory predicts demands much better in the more symmetric Nash–demand game. The highest entry is 45.3 which means that (γ^e, r) hit more than half of the observations on an area of 5%. Equity theory is worse in predicting acceptance bounds. But again, the top entries are higher for the Nash–demand game compared to the ultimatum game. One should also notice that (γ^r, r) is among the three best predictors is both games and is actually the best one in the ultimatum game.

Overall, equity standards (γ^s, r) and (γ^r, s) predict poorly. In addition, both are theoretically unreasonable. It is inconsistent to determine contributed shares to surplus and then ask for these very shares of the return. Alternatively, it does not make sense to calculate equitable shares according to contributions to return — i.e. without considering cost — and then ask for these share of the surplus — i.e. with taking cost into account. In any case the hit rate analysis is further evidence for the predictive power of equity theory.

In contrast, we did the same type of analysis for the game theoretic prediction and found that game theory organizes the data rather poorly, at least in the ultimatum game. Table 4.4 reports a small value for a_i^u and actually a negative one for d_i^u. The success measure is defined between −100 and +100 (for hit rate and area measured in percent). A negative value for m means that one would reach a higher success by predicting the complementary outcome space.[21] Although it requires equilibrium selection, the game theoretic solution fares well in the Nash–demand game[22]. But, one should notice that here it is equivalent to the equity standard (γ^e, r).

4.6.3 Equitable Decisions and Equitable Vectors

The hit rate analysis above looked at each of the two decisions taken in each game separately, but it does not say much about the distribution of equity types within the subject population. Our goal now is to evaluate whether or not the whole decision vector of player i is consistent with equity considerations and how frequently the different equity standards are applied, if demands, respectively acceptance bounds, are classified exclusively.

Each demand and acceptance bound (separately) is counted as an observation of equity standard (γ^k, l), if:

1. (γ^k, l) was the nearest neighbor out of the restricted set of equity standards, and
2. the distance of the observed d_i, respectively a_i, from the one predicted by (γ^k, l) was smaller than ±1.5 percentage points of the return.

We restricted the set of equity standards to be considered as nearest neighbors to $(\gamma^q, r), (\gamma^e, r), (\gamma^e, s)$ and (γ^r, r); i.e. to those standards that were rather successful according to the hit rate analysis. The cut–off ±1.5 percentage points — i.e. an area of 3% — is rather small. An increase in the cut–off increases the risk of falsely attributing behavior to equity considerations. Considering four equity standards as potentially relevant implies a combined prediction area of four times 3%. Table 4.5 reports the relative frequencies for the different equity standards. We combined the results for (γ^e, r) and (γ^e, s) — i.e. the results for "equal return splitting" and "equal surplus splitting". Whenever

[21] The bad performance of game theory in the ultimatum game replicates the findings in chapters 1 and 3.

[22] The entry for demand was left out, since the solution does not make a prediction for it.

Table 4.5. Relative Frequencies of Equitable Decisions

	Ultimatum Game		Nash–demand game	
	d_i^u	a_i^u	d_i^n	a_i^n
Equity				
(γ^q, r)	12%	5%	15%	4%
$(\gamma^e, r), (\gamma^e, s)$	32%	12%	24%	21%
(γ^r, r)	17%	15%	22%	17%
Unaccounted	39%	68%	39%	58%
$\sum$	100%	100%	100%	100%

either of the two was nearest neighbor they differed not at all or only marginally.[23]

According to table 4.5, more than 60% of the demands in both games can be classified as equitable. Egalitarian equity considerations are modal among the equity classes. All three categories attract substantial proportions of the data confirming our intuition that subjects exhibit heterogeneous equity types. Looking at the acceptance bounds, we find almost the same rank ordering, but at a lower level. Nonetheless, the proportion of acceptance decisions that are consistent with equity theory are 32% in the ultimatum game and 42% in the Nash–demand game. Overall, there are striking similarities between the two types of games. The stability of results is underlined by the fact that for 34% of the subjects either the demand or the acceptance bound or both was classified the same in both games.

Now we turn to evaluating the decision vectors. On the basis of the restricted nearest neighbor analysis presented above, we can classify vectors were both decisions are equitable, versus those were only one decision is equitable, versus others. We found no evidence for strategic demands in the ultimatum game, but there were some a^u–decisions matching the game theoretic solution. Furthermore, there were some rather low acceptance bounds, which were close to the players' own cost. We are interested in determining the frequency of such decisions, since they could be interpreted as behavior which is "anchored by strategic considerations".

The game theoretic solution in the U–Game predicts boundary decisions, but we also regarded extreme decisions — demands (acceptance bounds) above (below) a certain level — as consistent with strate-

[23] A few observations occurred were even the three remaining classes allowed no separation. Therefore the proportion of such mixed cases was assigned to the repective classes evenly.

Table 4.6. Relative Frequencies of Equitable Versus Extreme Decision Vectors

	All Data	First Game	Second Game	First Game Ult.	First Game Nash
Equity					
d_i and a_i	28%	18%	37%	18%	19%
only d_i	30%	34%	26%	27%	44%
only a_i	11%	13%	8%	9%	19%
Extreme d_i or a_i	8%	16%	0%	27%	0%
Unaccounted	23%	19%	29%	19%	19%
$\sum$	100%	100%	100%	100%	100%
# of obs.	38	38	38	22	16

gic reasoning in the U–Game. An acceptance bound was categorized as "extreme" if it was approximately equal to a player's own cost or below. Such an acceptance bound implies that the player gets a surplus share that is about zero or even negative. The qualifier "approximately" meant that $a_i \approx q_i$ implied a difference less than ± 1.5 percentage points of the return. A demand was categorized as "extreme" if it was approximately equal to or higher than return minus the opponent's cost. Such a demand meant that a player asked for his cost and all surplus or more.

Table 4.6 shows the relative frequencies of the respective vector choices. The result for all data says that 28% of all vectors are such that both components are equitable. In addition 30% of the vectors exhibit an equitable demand, but no equitable acceptance bound. And another 11% of the vectors are such that a_i is equitable, but not d_i. Thus, in total, 69% of all decision vectors can be characterized as equitable.

Looking at the data for the first and the second game separately, we find that the proportion of equitable vectors increases from 65% to 71%. Among the equitable vectors, there is a shift toward those were both components are equitable. This is clear evidence for what we call "purely equitable behavior". Almost half of all vectors in the first game and more than one third in the second game indicate equity anchoring, since one component is set at an equitable level while the other is not. Extreme vectors were observed only in the first game — to be exact, only if the ultimatum game was played first. In the latter case, these vectors exhibit a substantial proportion of 27%. However, since this evidence vanishes with experience, it is only a weak indication of strategic reasoning in the ultimatum game.

There are more cases in which only the demand is equitable than cases in which only the acceptance bound is equitable. Thus, it seems that subjects regard the demand as the target variable for implementing an equitable allocation. This is quite plausible in the ultimatum game, since the pie is divided as specified by the demand. The acceptance bound determines only whether an agreement is reached or not. But in the Nash–demand game, in case of an A-agreement, the division depends on both acceptance bounds. Remember that above we described two different kinds of equity anchoring: "equitable d_i and undercutting" versus "equitable a_i and overbidding". We find evidence for both, but far more subjects can be characterized by the first type of equity anchoring rather than the second one.

4.6.4 A Further Look at Treatment Effects

We can now investigate in more detail the influences of asymmetric productivities and strategic asymmetry. The regression analyses above proved the overall influence of individual productivities. Investigating individual decisions in detail we found more specifically, that asymmetries in players' productivities play a role at least for those subjects, who follow in one or both of their decisions the equity standard (γ^r, r). These are about 20% of the players in both types of games (see table 4.5).

Strategic asymmetry should not influence the behavior of subjects which are purely equity motivated. But one should expect an effect upon the tactical bargaining margin, if decisions are equity anchored. Whereas the anchor is not influenced by strategic aspects of the game, the tactical decision is. Specifically, overbidding an equitable a_i and undercutting an equitable d_i should be stronger in the U–Game. Therefore we defined the tactical bargaining margin of player i (TBM_i) empirically as

$$TBM_i \equiv \frac{d_i}{r} - \frac{a_i}{r}.$$

That is, TBM_i is the difference between the return share demanded and the one chosen as acceptance bound.[24] We tested the following hypothesis:

[24] This is a simple operationalization of the notion of a tactical bargaining margin. It might be more reasonable to use one based on the demanded and acceptable surplus share, but we leave this for further investigations.

For equity anchored subjects the tactical bargaining margin is greater in the ultimatum game than in the Nash–demand game!

We used first game data to test the hypothesis. The mean of TBM_i is 0.20 in the U–Game and 0.12 in the N–Game. A Mann–Whitney U–Test confirms our hypothesis ($\alpha < 0.05$, one–tailed). Running an equivalent test for subjects not classified as equity anchored, we found no significant difference between the two games.[25]

4.7 Summary

"Equity anchoring" is a model for bargaining behavior. The format of bargaining decisions collected in our experiment allows us to separate between a subject's equity target and its tactical bargaining margin, and we classified the decision as well as the decision vectors. The empirical findings are:

Result 4.5 *The subject population exhibits heterogenous equity types. There are three types which are present in substantial parts of the population:*

— (γ^q, r) : *dividing return proportional to cost (neglecting productivities)*
— (γ^r, r) : *dividing return proportional to contributions to return (considering productivities, but neglecting cost)*
— $(\gamma^e, r), (\gamma^e, s)$: *dividing return or surplus equally*

Result 4.6 *About 70% of the decision vectors are equitable in the sense that at least one component hits an equitable allocation within a distance of ±1.5% of the available return:*

— *around 30% are purely equitable (both components)*
— *around 40% are equity anchored (only one component).*

Result 4.7 *Demand is the dominant decision variable to implement an equity target; i.e. vectors of type "equitable demand and undercutting" are observed more often than those of type "equitable acceptance bound and overbidding".*

Result 4.8 *We find more cases of equitable behavior and a reduced tactical bargaining margin in the Nash–demand game than in the ultimatum game.*

[25] We neither found significant differences for MWU nor the Kolmogorov–Smirnov 2–Sample Test.

Result 4.9 *The strategic model of ultimatum bargaining behavior receives little support. The cooperative Nash–Solution fares better in explaining the data, but nevertheless it seems inferior to equity theory.*

Overall, we found strong evidence for equity concerns in bargaining games with advance production confirming findings in similar games.[26] Furthermore, our study determines distributions of equity types according to the decisions as well as the decision vectors.

Result 4.8 may be seen to suggest that the size of the TBM is chosen with considering the strategic solution of the game. While we cannot rule out such an explanation, it is not the one we favor. We rather prefer a qualitative interpretation: A first mover (second mover) in an ultimatum may feel superior (inferior) even though he never determines and consequently never considers the perfect equilibrium.

It is interesting to speculate about the consequences of equity anchoring in a wider context. In a dynamic bargaining game a player's demand at some point in time may, for instance, comprise two components: the equity target he intends to reach and the TBM. Within the bargaining process a player may adjust his TBM, but one can also imagine that his equity anchor shifts. Especially in bargaining with incomplete information (see e.g. Mitzkewitz and Nagel, 1993) the latter could be an important consequence of communication or signalling. It could mean that communication is valuable and might offer an explanation for costly delays.

---

[26] See Berg, Dickhaut and McCabe (1995), Fehr, Gächter and Kirchsteiger (1997), Güth, Klose, Königstein and Schwalbach (1998), Hackett (1993), Königstein (1998a, see chapter 3) and Königstein and Tietz (1998, see chapter 1).

5. Efficiency and Evolution of Social Preferences and Prosocial Behavior

5.1 Introduction

One might view social life as a sequence of competitive markets were a large number of economic agents interact anonymously and were individual decisions have only a negligible influence on each other. This is certainly an extreme description of the world. At the other extreme one might model social life as a sequence of two person games in which both players know each other very well and were each individual's well–being depends to a large extent on the other player's decisions. Of course, the truth will be a mixture of both and there are other intermediate forms of interaction, e.g. small group interaction, as well. The analysis here relies on a two–player model. It will be shown that in this case social preferences of the agents can generate economic efficiency and, moreover, that such preferences can be evolutionarily stable. Since social preferences may generate prosocial actions the study is at the same time on the evolution of prosocial behavior.

Investigating the consequences of evolution for economic problems is not novel. An early example is Alchian (1950) who showed that environmental selection can replace rational profit–maximizing behavior. Social preferences are modelled as an altruistic utility function, i.e., as a utility function in which an agent's utility is a weighted average of both players' payoffs.[1] As a polar case, this specification allows also for egoistic preferences — which is the standard assumption in economics. It is shown that egoism may be driven out by altruistic

[1] As a source of motivation for studying the evolution of altruism, I want to refer to a statement by Hirshleifer (1977, p. 19): "In any attempt to broaden the application of economic reasoning, to make it a general social science, a key issue is the problem of altruism (the 'taste' for helping others): its extent, provenance and determinants."

preference types in the process of evolution. The reason is that egoism may generate inefficiency which can be avoided by altruistic players.

Imagine that once in a while you see an opportunity to undertake a joint project with some partner. E.g. you can organize a scientific conference with a colleague, coach a sports team with another trainer or manage a company together with your business partner. All three case are examples for social production and it is not hard to imagine that whether or not you join the project will depend on how the return is distributed afterwards.

Even though it may sometimes be hard to specify what the return is — e.g. in the first two examples it might be social status —, you look at your partner and you know one thing: whatever the returns are, he will try to get away with most of it. So you decide better not to join, but you would have, if the other were less greedy.

In the business case you might offer your partner a contract. If he bargains for too much, you may decide to run your business alone or you may be better off by employing your partner and not working yourself. By avoiding your own cost of work effort this might be profitable. All three cases illustrate how egoism may generate inefficiency in production. The formal model introduced below captures these ideas.

More generally, a socio–economic conflict can be viewed as consisting of two parts: a social production process and a distribution conflict. In the distribution of a given resource, egoistic players will capture higher shares, than altruists. But, since receiving a smaller share in distribution may deteriorate the incentives to produce the resource in the first place, egoism may lead to inefficiency in production. The loss due to this inefficiency may outweigh the advantage of egoists due to capturing higher shares. Furthermore, if the development of individual preferences follows an evolutionary process in which the reproduction of a certain preference type depends on the material returns it generates, one can easily imagine that non–egoistic preferences are superior to egoistic ones. To check this intuition in a formal model, however, is not as simple as it may seem. Efficiency gains are not enough: For evolutionary stability it is necessary that a population of agents, who are endowed with certain preferences, must be stable against every small invasion of potential entrants. One can imagine that an egoist entering a society of altruists could even better exploit his partners.

These arguments are loose and intended to generate a more general view of the theme. In the formal model below the agents play exactly one game during their lifetime. Each game consists of a bargaining part in which the two players can agree upon the split of a return to be produced subsequently. Production decisions are 0–1–decisions, i.e. either to work or not. To work causes individual cost and by means of two cost parameters social production can be more or less efficient. Evolutionary success of a certain preference type depends on the economic profit received by the agent who is endowed with these preferences.

The agents are assumed as being rational with respect to their individual preferences. They evaluate the game outcomes according to their utility functions, and this is common knowledge. Accordingly, they play a subgame perfect equilibrium of the game. However, maximizing utility is not equivalent to maximizing evolutionary success. By evaluating the equilibrium actions according to their economic profitability the result of evolution is studied within the concept of evolutionary stability.

This procedure, of generating actions by assuming some individual preferences and thereafter evaluating them according to a possibly different evolutionary success function, is the heart of the "indirect evolutionary approach" which was invented by Güth and Yaari (1992).[2] For a general description of this method see Königstein and Müller (forthcoming).

By choosing simple features the formal model gets tractable, but of course, it loses realism. E.g. one might think that common knowledge of the players' preference types is unrealistic. Such criticism is valid, but on the other hand assuming that in two-person interaction the agents have no idea at all about each others' preferences seems even more unrealistic. Some shortcomings of the model will be discussed below, but after all this is just one step to explain the evolution of social preferences and prosocial behavior.

Finally, one may doubt whether concern for others does exist at all. Besides the evidence on charity spending those readers shall be directed to laboratory experiments.[3] A lot of experimental results give rise to speculations about social preferences or internalized social norms — which may affect behavior the same way as preferences.

[2] There is a growing literature applying this approach; see e.g. Bester and Güth (1998), Dufwenberg and Güth (1999), Güth (1995a), Güth and Kliemt (1994), Güth and Peleg (1997), Huck (1998) and Huck and Oechsler (1999).

[3] See e.g. the recent survey on bargaining experiments of Roth (1995).

In fact, to combine an ultimatum game with a production game within the theoretical study here, was partially motivated by the experiments reported in chapters 1, 3 and 4. In these experiments, subjects exhibit behavior which is inconsistent with that of egoistic and purely money motivated agents. They seemed to care for fair allocations and what was considered fair was related to their productive effort. In the theoretical study here, agents are not modelled as being endowed with a preference for equity, but nevertheless the model can explain why moderate, compared to egoistic, bargaining demands occur. It does not allow for disagreement equilibria. However, it will be indicated below that a stochastic extension of the model — which is left for future research — may yield such equilibria. So, the spirit of the experiments — namely, to study the relationship between distribution and production decisions — carries over to the theoretical analysis here, but it will not be tried to directly relate the theoretical results with the empirical data.

In the following, first, I will describe the ultimatum game with social production and derive its game theoretic solution. Then, the indirect evolutionary game will be described, which incorporates the ultimatum game with social production as base game. Finally, the indirect evolutionary game will be solved for generic parameter constellations, relying on the concept of evolutionary stability.

5.2 Ultimatum Bargaining with Subsequent Social Production

Consider two players, 1 and 2, who can agree on how to share a joint output (monetary return) which is determined afterwards by simultaneous production choices. Specifically, the game is described by the following three stages:

1. Player 1 chooses $s_1 \in [0,1]$ and informs player 2 about his choice. The variable s_1 represents the share of the return player 1 demands for himself.
2. Player 2 decides whether to accept ($\delta_2 = 1$) or reject ($\delta_2 = 0$) the demand and informs player 1 about his choice.
3. Each player chooses his individual input level $q_i \in \{0,1\}$, $i = 1,2$. Production choices are made simultaneously. This ends the game.

Production causes individual cost $c_i(q_i) \geq 0$ with

$$c_i(q_i) = \begin{cases} c_i \text{ if } q_i = 1 \\ 0 \text{ otherwise} \end{cases}$$

and determines return R according to the following social production function

$$R(q_1, q_2) = q_1 + q_2.$$

So, the players monetary payoffs (profits) are determined according to the following profit functions:

$$\Pi_1(s_1, \delta_2, q_1, q_2) = \delta_2 s_1 R(q_1, q_2) - c_1(q_1)$$
$$\Pi_2(s_1, \delta_2, q_1, q_2) = \delta_2 (1 - s_1) R(q_1, q_2) - c_2(q_2)$$

The bargaining procedure described above is known as "ultimatum game" (see Güth, Schmittberger and Schwarze (1982)). It is a rather asymmetric bargaining game in which the first mover has a strong strategic advantage. Player 2 faces a "take-it-or-leave-it" situation, and therefore he should accept 1's demand, even if he gets nothing but his reservation value (see the game theoretic analysis below). Strategic asymmetry in bargaining may reflect asymmetric power or asymmetric property rights, and for obvious reasons, player 1 will therefore be referred to as "firm owner" and player 2 as "worker".

That input may be zero or one can be thought as choosing a low or high effort level or to shirk or work. For the moment, simultaneous production choices will be assumed. The theoretical analysis will reveal that this makes no difference to the case of sequential choices. The marginal productivity of high effort is normalized to 1 for each player. Cost are normalized to zero in case of low effort and are non–negative in case of high effort. Cost may differ between firm owner and worker and they may be greater or lower than i's marginal productivity. This reflects that, in general, social production is asymmetric. Either productivities or cost or both vary across individuals. Normalizing both productivities to 1 and allowing the cost parameters c_1 and c_2 to vary freely is a convenient way to represent these potential asymmetries, since it requires only two parameters. A vector (c_1, c_2) represents a certain social production technology. It will be referred to as "technology".

While monetary payoffs are given by $\Pi_1(.)$ and $\Pi_2(.)$, the players have preferences on social allocations, which shall be represented by the following utility functions:

$$U_i(s_1,\delta_2,q_1,q_2) = t_i\Pi_i(.) + (1-t_i)\Pi_j(.)$$
$$\forall(i,j) \in \{(1,2),(2,1)\}$$

and with

$$t_i \in \left[\tfrac{1}{2},1\right].$$

The parameter t_i represents player i's preference type which, in general, differs from player j's type. It will be assumed that the preference types are common knowledge.

So, the players derive utility not only from their own income, but also from their opponent's income. Utility is a weighted average of both $\Pi_1(.)$ and $\Pi_2(.)$. If $t_i = \frac{1}{2}$, this means that the respective player is an extreme altruist. He does not care whether he himself or the other player gets the monetary payoff; his marginal utility of economic wealth is independent of its distribution within the society. If $t_i > \frac{1}{2}$, a player's marginal utility of income is higher, if income is given to himself rather than to the other player. This exhibits a nice feature of the model which certainly matches one's intuition: Except for the degenerate case with $t_i = \frac{1}{2}$, all types of players are at least moderately egoistic. Alternatively, one might call these types moderately altruistic. In general, to avoid confusion the terms "egoist" and "egoism" will be reserved for the extreme case, i.e., $t_i = 1$, while cases $t_i < 1$ will be referred to as more or less altruistic.

To describe social preferences by this type of utility function is common. It was used by Bester and Güth (1998), and they refer to Walras, who had already used it. However, there is no general agreement on how to formulate altruism. For another definition see e.g. Becker (1976) and Loubergé and Schlesinger (1988).[4] Afterall, the labels "altruistic" and "egoistic" are used here simply to identify subclasses of preferences.

5.3 Game Theoretic Solution

Let X_1, respectively X_2 denote the pure strategy set of player 1, respectively player 2, in the ultimatum game with subsequent produc-

[4] The latter formalize egoism and altruism within a simple bargaining game. According to their view an individual whose marginal utility of own income is greater than that derived from the other player's consumption — which is the case here — is still called an egoist.

tion as described above. So, formally the game is defined by the tuple $(X_1, X_2, U_1(.), U_2(.))$. For reasons that will become clear later on, this game will be referred to as "base game" $G^B(t_1, t_2)$. It depends among other things on the commonly known preference types t_1 and t_2. The evolutionary analysis which follows in later sections relies on the game theoretic solution of $G^B(t_1, t_2)$ introduced here. The solution of $G^B(t_1, t_2)$ will be derived according to the notion of subgame perfect equilibrium (SPE).[5] To simplify notation let α, respectively β, represent the preference parameter of player 1, respectively player 2. Thus the utility functions can be written as:

$$\begin{aligned} U_1(.) &= \alpha \Pi_1(.) + (1-\alpha)\,\Pi_2(.) \\ &= \alpha\left[\delta_2 s_1 (q_1 + q_2) - c_1(q_1)\right] \\ &\quad + (1-\alpha)\left[\delta_2 (1 - s_1)(q_1 + q_2) - c_2(q_2)\right] \qquad (5.1) \\ U_2(.) &= (1-\beta)\,\Pi_1(.) + \beta \Pi_2(.) \qquad (5.2) \end{aligned}$$

The model has four parameters, α, β, c_1 and c_2, and in the following a solution will be determined for all vectors $(\alpha, \beta, c_1, c_2) \in \left(\frac{1}{2}, 1\right]^2 \times R_+^2$. The solution for cases $\alpha = \frac{1}{2}$ and/or $\beta = \frac{1}{2}$ will be characterized thereafter.

The game has three stages. An SPE induces an equilibrium in every subgame. So, the solution will be determined by working backwards from the last to the first stage.

5.3.1 Stage 3: Production

The third stage of the game consists of the production decisions by both players. If $\delta_2 = 0$ — i.e. bargaining resulted in disagreement — no social production takes place. Both players receive zero utility which can be considered an outside option value. Let $U_i(q_1, q_2)$ be a shorthand for $U_i(q_1, q_2 \mid s_1, \delta_2)$. Thus if $\delta_2 = 1$ the game at stage 3 can be described by the normal forms displayed in table 5.1.

Assumption 5.1 *If player i is indifferent at stage 3 between high and low effort, then $q_i = 1$.*

As one can easily see, player 1 should choose high effort $q_1 = 1$ if and only if

[5] See Selten (1965) and (1975).

Table 5.1. The Game at Stage 3

	$q_2 = 0$	$q_2 = 1$
$q_1 = 0$	$U_1(0,0)$; $U_2(0,0)$	$U_1(0,1)$; $U_2(0,1)$
$q_1 = 1$	$U_1(1,0)$; $U_2(1,0)$	$U_1(1,1)$; $U_2(1,1)$

Using equations 5.1 and 5.2 one gets:

	$q_2 = 0$	$q_2 = 1$
$q_1 = 0$	0; 0	$\alpha s_1+(1-\alpha)((1-s_1)-c_2)$; $(1-\beta)s_1+\beta((1-s_1)-c_2)$
$q_1 = 1$	$\alpha(s_1-c_1)+(1-\alpha)(1-s_1)$; $(1-\beta)(s_1-c_1)+\beta(1-s_1)$	$\alpha(s_1 2-c_1)+(1-\alpha)((1-s_1)2-c_2)$; $(1-\beta)(s_1 2-c_1)+\beta((1-s_1)2-c_2)$

$$s_1 \geq \underline{s}_1 \equiv \frac{\alpha(1+c_1)-1}{2\alpha-1} \tag{5.3}$$

Condition 5.3 is an incentive compatibility constraint for player 1. Notice:

- If $c_1 > 1$, there exists no $\alpha \in \left[\frac{1}{2}, 1\right]$ such that 5.3 is satisfied.
- If $\alpha < \frac{1}{1+c_1}$, 5.3 cannot be binding.

The former would require $s_1 > 1$ which is not feasible while the latter implies $s_1 \geq 0$ which is always true.

Accordingly, player 2 chooses high effort $q_2 = 1$ if and only if

$$s_1 \leq \bar{s}_1 \equiv \frac{\beta(1-c_2)}{2\beta-1} \tag{5.4}$$

This is the incentive constraint for player 2. Note:

- If $c_2 > 1$, there exists no $\beta \in \left[\frac{1}{2}, 1\right]$ such that 5.4 is satisfied.
- If $\beta < \frac{1}{1+c_2}$, 5.4 cannot be binding.

The former would require $s_1 < 0$ which is not feasible while the latter implies $s_1 \leq 1$ which is always true. Equilibrium in the subgame starting at stage 3 is therefore characterized by

$$q_1^* = 1 \text{ if and only if } \delta_2 = 1 \text{ and 5.3 holds,}$$

$$q_2^* = 1 \text{ if and only if } \delta_2 = 1 \text{ and 5.4 holds.}$$

This implies unique play in the subgame starting at stage 3.

5.3.2 Stage 2: Acceptance or Rejection

At stage 2, player 2 decides upon acceptance ($\delta_2 = 1$) or rejection ($\delta_2 = 0$) of 1's demand. In case of rejection, no work contract is agreed upon and no joint production takes place subsequently. Both players' profits are zero and accordingly their utilities are zero.

Assumption 5.2 *If player 2 is indifferent between acceptance and rejection, then $\delta_2 = 1$.*

Given rational play in the subgame starting at stage 3, one finds that $U_2(\delta_2 = 1 \mid s_1, q_1^*, q_2^*) \geq 0$ for all constellations of the two incentive constraints and therefore $\delta_2^* = 1$ in every SPE.

5.3.3 Stage 1: Demand

At stage 1, player 1 chooses s_1 in order to maximize $U_1(s_1 \mid \delta_2^*, q_1^*, q_2^*)$. Intuitively, an increase in s_1 has two effects. First, for fixed values of $\bar{\delta}_2, \bar{q}_1$ and $\bar{q}_2$, it shifts return from player 2 to player 1. Secondly, the choice of s_1 might alter subsequent choices (δ_2, q_1 and q_2) according to 5.3 and 5.4. For $\bar{\delta}_2 = 1$, the first effect is captured by

$$\frac{\partial U_1(s_1 \mid \bar{\delta}_2, \bar{q}_1, \bar{q}_2)}{\partial s_1} = \alpha \bar{R} - (1-\alpha)\,\bar{R} \tag{5.5}$$

which is non–negative given the restrictions of the model.[6] According to 5.5 player 1 should choose s_1 as high as possible, as long as subsequent decisions remain unchanged. However, since subsequent play depends on s_1, it might be advantageous to choose s_1 less than maximal. Specifically, a small demand can be advantageous, if it induces player 2 to choose high instead of low effort. Since $s_1 = \bar{s}_1$ is sufficient for $q_2^* = 1$, it follows that in every SPE with $q_2^* = 1$

$$s_1^* = \begin{cases} 1 \text{ if } \beta < \frac{1}{1+c_2} \\ \bar{s}_1 \text{ otherwise.} \end{cases}$$

[6] If $R = 0 \iff q_1 = q_2 = 0$, the choice of $s_1$does not matter. This case is not interesting, since there is nothing to be distributed.

5.3.4 Subgame Perfect Equilibria

Now it is possible to characterize the solution for all $(\alpha, \beta, c_1, c_2) \in \left(\frac{1}{2}, 1\right]^2 \times R_+^2$. It's clear from above that in every SPE

$$\begin{aligned} q_1^* &= 1 \text{ if and only if 5.3 holds,} \\ q_2^* &= 1 \text{ if and only if 5.4 holds and} \\ \delta_2^* &= 1. \end{aligned}$$

Let $\sigma = (s_1, \delta_2, q_1, q_2)$ be a shorthand for a strategy profile of the game $G^B(t_1, t_2)$ and σ^* a subgame perfect equilibrium. Note, that σ is not a strategy profile in a proper game theoretic sense. Namely, it describes only a single path of play. Accordingly, σ^* describes an SPE–path. A complete description of a strategy profile would require to specify play at all informations sets. However, for all analyses that follow it suffices to specify the path of play. Therefore using σ as a shorthand for a strategy profile will be convenient. Furthermore, let σ_{kl}, with $k, l \in \{0, 1\}$, denote a strategy profile with $q_1 = k$, $q_2 = l$, and let σ_{kl}^* denote an according SPE (i.e., $q_1^* = k$, $q_2^* = l$, $\delta_2^* = 1$) in which s_1^* is left unspecified. Thus, an SPE can be described by σ_{kl}^* and a specification for s_1^*.

In some cases the game theoretic solution is trivial. This is the content of the following proposition.

Proposition 5.1 *For $c_i > 1$ and/or $\beta < \frac{1}{1+c_2}$ the game theoretic solution of the base game is:*

$$\begin{aligned} &\sigma_{00}^*, s_1^* \in [0, 1] \\ &\quad \textit{if } c_i > 1 \textit{ for both players,} \\ &\sigma_{01}^*, s_1^* = \min\{\bar{s}_1, 1\} \\ &\quad \textit{if } c_1 > 1,\ c_2 \leq 1, \\ &\sigma_{10}^*, s_1^* = 1 \\ &\quad \textit{if } c_2 > 1,\ c_1 \leq 1 \textit{ or} \\ &\quad \textit{if } c_i \leq 1 \textit{ for both players and } \beta < \frac{1}{1+c_2}. \end{aligned}$$

Proof. It was shown above that $c_i > 1 \Rightarrow q_i = 0$ and that $s_1 = 1$ satisfies 5.4, if $\beta < \frac{1}{1+c_2}$. Combining these conditions yields the result.

Proposition 5.1 characterizes the solution for trivial specifications of the parameters. Non–trivial cases are such that (c_1, c_2) lies in the unit square and player 2 is not too altruistic. Then the following definitions and conditions are important.

Definition 5.1 *Let*

$$\Delta S_1 \equiv \underline{s}_1 - \bar{s}_1,$$
$$\Delta U_1^1 \equiv U_1(1,0) \mid_{s_1=1} - U_1(0,1) \mid_{s_1=\bar{s}_1} \quad \textit{and}$$
$$\Delta U_1^2 \equiv U_1(1,0) \mid_{s_1=1} - U_1(1,1) \mid_{s_1=\bar{s}_1} .$$

$U_1(0,1) \mid_{s_1=\bar{s}_1}$ is the utiliy player 1 recieves if $q_1 = 0$, $q_2 = 1$, $\delta_2 = 1$ and $s_1 = \bar{s}_1$. $U_1(1,0) \mid_{s_1=1}$ and $U_1(1,1) \mid_{s_1=\bar{s}_1}$ are defined analogously. In case

$$\Delta S_1 > 0 \Longleftrightarrow \underline{s}_1 > \bar{s}_1,$$

only one of the incentive constraints can be satisfied, but not both. Thus, by choosing s_1, player 1 can induce either σ_{10} or σ_{01}— given subsequent equilibrium decisions. His choice of s_1^* depends on the sign of ΔU_1^1. I.e., it depends on a comparison of $U_1(1,0) \mid_{s_1=1}$ and $U_1(0,1) \mid_{s_1=\bar{s}_1}$, which are respective utilities of player 1 conditional on the choice of s_1. If $\Delta U_1^1 \leq 0$ he should induce σ_{01}^*, otherwise he should induce σ_{10}^*.

ΔU_1^2 is defined equivalently. If

$$\Delta S_1 \leq 0 \Longleftrightarrow \underline{s}_1 \leq \bar{s}_1,$$

both incentive constraints can be satisfied — i.e. player 1 will choose $q_1 = 1$ in every case (given other equilibrium choices). Therefore player 1 should induce σ_{11}^*, if $\Delta U_1^2 \leq 0$, and σ_{10}^* otherwise.

ΔS_1, ΔU_1^1 and ΔU_1^2 are functions of the cost parameters and the players' types. They play a crucial role throughout the analysis, since the signs of ΔS_1, ΔU_1^1 and ΔU_1^2 determine which production vector is reached in equilibrium. Within the indirect evolutionary game below they also determine which production is feasible via mutation. Cases of indifference, $\Delta S_1 = 0$, $\Delta U_1^1 = 0$ or $\Delta U_1^2 = 0$, will be solved according to the following assumption.

Assumption 5.3

If $\Delta S_1 > 0$ *and* $\Delta U_1^1 \leq 0$, *then* σ_{01}^*.
If $\Delta S_1 \leq 0$ *and* $\Delta U_1^2 \leq 0$, *then* σ_{11}^*.
Otherwise, play is σ_{10}^*.

Combining the above results and definitions proves the following proposition.

Proposition 5.2 *For $c_1 \leq 1$, $c_2 \leq 1$ and $\beta \geq \frac{1}{1+c_2}$ the game theoretic solution of the base game is:*

$$\begin{array}{ll} \sigma_{01}^*, s_1^* = \bar{s}_1 & \text{if } \Delta S_1 > 0 \wedge \Delta U_1^1 \leq 0 \\ \sigma_{11}^*, s_1^* = \bar{s}_1 & \text{if } \Delta S_1 \leq 0 \wedge \Delta U_1^2 \leq 0 \\ \sigma_{10}^*, s_1^* = 1 & \text{otherwise.} \end{array}$$

This is the game theoretic solution for all non–trivial constellations of the parameters. For all $(\alpha, \beta, c_1, c_2) \in \Omega^4$ with

$$\Omega^4 \equiv \left\{ (\alpha, \beta, c_1, c_2) \mid (\alpha, \beta) \in \left(\frac{1}{2}, 1\right]^2 \wedge (c_1, c_2) \in [0, 1]^2 \right\}$$

— this is the relevant parameter space for the evolutionary analyses below — proposition 5.2 essentially characterizes a unique solution of the game $G^B(t_1, t_2)$. The attribute "essentially" is supposed to indicate that there are cases of indifference which are non–generic and which were solved by assumption. Actually, some of the assumptions are not necessary. For instance that player 2 chooses $q_2 = 1$ instead of $q_2 = 0$ when his incentive constraint is binding is an implication of subgame perfect equilibrium (for non–trivial parameter constellations). However, not all indifferences can be resolved that way. E.g. if $\Delta S_1 \leq 0$ and $\Delta U_1^2 = 0$, then player 1 could also induce σ_{10}^* instead of σ_{11}^*. So, the above assumptions are mainly for convenience; one does not have to wonder about paramter constellations that are trivial or exhibit zero measure relative to Ω^4.

It remains to solve cases $\alpha = \frac{1}{2}$ and/or $\beta = \frac{1}{2}$. In case a player is an extreme altruist his effort is independent of s_1 and depends solely on the technology. Specifically, if $\alpha = \frac{1}{2}$ $\left(\beta = \frac{1}{2}\right)$, player 1 (2) should choose $q_1^* = 1$ $(q_2^* = 1)$ whenever $c_1 \leq 1$ $(c_2 \leq 1)$. Given this production behavior player 2 should choose $\delta_2^* = 1$, and it only remains to determine s_1^* which is

$$\begin{array}{ll} s_1^* = 1 & \text{if } \alpha > \frac{1}{2} \wedge \beta = \frac{1}{2} \\ s_1^* \in [0, \bar{s}_1] & \text{if } \alpha = \frac{1}{2} \wedge \beta > \frac{1}{2} \\ s_1^* \in [0, 1] & \text{if } \alpha = \frac{1}{2} \wedge \beta = \frac{1}{2}. \end{array}$$

5.3.5 Properties of the Game Theoretic Solution

As shown, depending on the parameters one of the following types of SPE is reached (for non–trivial parameter specifications):

- The σ^*_{10}–equilibrium: the firm owner demands all return for himself $s^*_1 = 1$ and works alone.
- The σ^*_{01}–equilibrium: the employee works alone. The firm owner demands $s^*_1 = \bar{s}_1$, leaving the worker an incentive compatible share of the return.
- The σ^*_{11}–equilibrium: both players choose to work and the firm owner demands $s^*_1 = \bar{s}_1$.

Given (c_1, c_2) within the unit square, only σ^*_{11} is economically efficient — i.e. generating the maximal sum of profits.[7] Both other equilibria are inefficient. A central result, which will be rather important for the evolutionary analysis, is that altruism may enhance economic efficiency. Part of this is established by the following proposition.

Proposition 5.3 *The support of σ^*_{11} within the unit square of the technology space is decreasing in egoism, if*

player types are equal, i.e. $\alpha = \beta$ or
player 2 is not too altruistic in the sense that $\beta > \frac{3-c_2}{2(1+c_2)}$.

Proof. For all $(c_1, c_2) \in [0,1)^2$ and $(\alpha, \beta) \in \left(\frac{1}{2}, 1\right]^2$:

$$\begin{array}{ll} \Delta S_1(\alpha,\beta) & \text{is strictly increasing in } \alpha \text{ and } \beta, \\ \Delta U_1^2(\alpha,\alpha) & \text{is strictly increasing in } \alpha \\ \Delta U_1^2(\alpha,\beta) & \text{is strictly increasing in } \beta \text{ and} \\ & \text{strictly increasing in } \alpha \text{ if } \beta > \dfrac{3-c_2}{2(1+c_2)}. \end{array}$$

Proposition 5.3 is illustrated in figures 5.1 to 5.3.

Given $\alpha = \beta = 1$, figure 5.1 displays the supports of the different types of equilibria within the unit square of the technology space.

[7] Efficiency can also be measured with respect to utilities. But throughout the analyis the term "efficiency" means "economic efficiency". This is consistent with the pursued goal, which is to study the economic consequences of altruistic preferences in the short run and their survival in the long run.

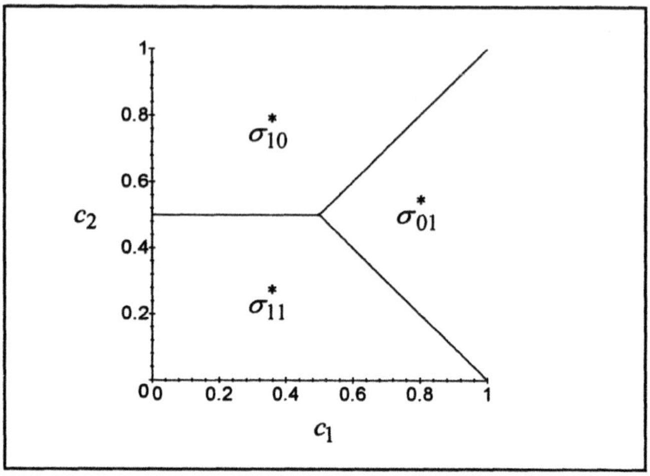

Fig. 5.1. SPE–Regions for $\alpha = \beta = 1$.

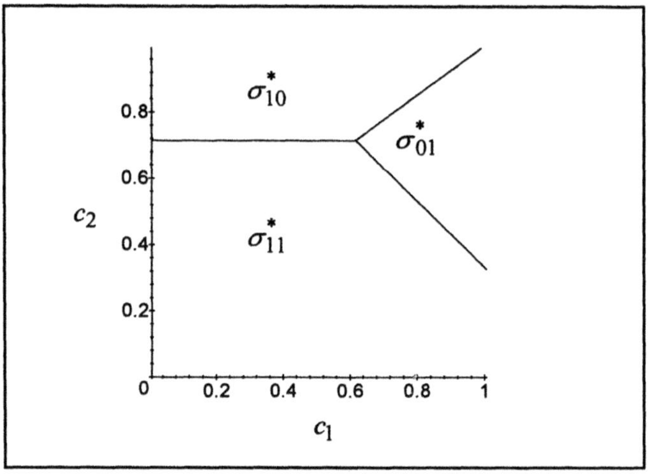

Fig. 5.2. SPE–Regions for $\alpha = \beta = \frac{3}{4}$.

Figure 5.2 shows analogous regions for $\alpha = \beta = \frac{3}{4}$ and figure 5.3 superimposes the former two.

In interpreting the results, let's assume for the moment that one observes two populations of players (societies), where each individual is endowed with social preferences as described here. They play the game pairwisely and the technology differs for different pairs, but is equal on average for the two societies. According to proposition 5.3

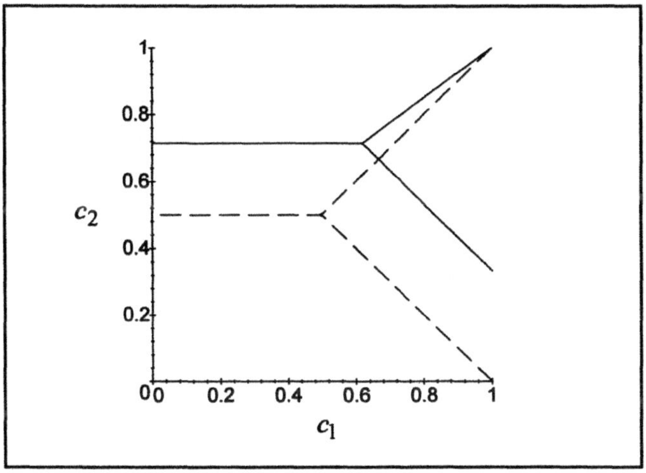

Fig. 5.3. Display of Both Previous Figures Together.

altruism can increase income; or — to choose a more political interpretation: Egoistic societies may be economically less efficient than socially concerned societies.

In the beginning it was said that this study investigates the economic consequences and the evolution of social preferences and prosocial behavior. It was not said clearly, what prosocial behavior is, but this shall be done now. Behavior is assumed to be equilibrium behavior and having derived the game theoretic solution, one can now look for the connection between preferences and equilibrium behavior. Everything else equal, the following holds:

1. Egoistic workers require higher bargaining shares $(1 - s_1^*)$ to choose "work" instead of "not work".
2. Egoistic firm owners choose joint production less often — i.e., the technology set that supports efficient production is smaller. Compared to altruistic firm owners they decide more often to work alone (σ_{10}^*) or to extract only their bargaining share, while letting the employee work alone (σ_{01}^*).

Given an equilibrium with $q_2^* = 1$, statement 1 says that the worker's behavior directly effects the distribution of income between worker and employer. But, egoism on the workers' side has also an indirect effect, since it is anticipated by the firm owners and may change the type of equilibrium. Joint production is reached less often such that egoistic workers not only influence the distribution of income within the society, but also reduce its total amount. Similar arguments can be given for the case of egoistic firm owners (statement 2).

Accordingly, within this model prosocial behavior shall be defined in the following relative sense: Given the same bargaining shares, players who choose higher work effort show prosocial behavior, as well as firm owner's who decide more often for joint production. Furthermore, players who require a smaller bargaining share to respond a certain work effort exhibit prosocial behavior.[8]

The solution was also derived for $c_i > 1$ to investigate the implications of altruistic preferences in cases were the net productivity of player i is negative. In equilibrium social preferences do not lead to economically inefficient production — since $c_i > 1$ implies $q_i^* = 0$. This is certainly a nice regularity of the model. Furthermore, the model allows for a multiplicity of observable equilibrium demands s_1^*, which matches a stylized fact of experimental bargaining behavior.

After all, that social preferences can generate economic efficiency may not be too surprising. For instance, imagine a prisoner's dilemma game in which the payoffs given in the description of the game represent money rather than utility. If two altruists — i.e. players who are endowed with a social preference function as defined above and $t_i < 1$ — play this game and if their concern for the other player's monetary payoff is sufficiently high, then each player may choose the cooperative action thereby reaching the efficient outcome. An important question is, however, whether such behavior is likely to exist in the long run. For the case of the ultimatum game with subsequent production rather than the prisoner's dilemma, this is in fact the question investigated below. Therefore, within the following indirect evolutionary game Γ (to be defined below) the base game $G^B(t_1, t_2)$ is embedded in an evolutionary framework.

[8] A more general definition of prosocial behavior can be found in Lück (1975), who empirically investigates helping behavior. He considers every *positive* form of social behavior as "prosocial behavior".

5.4 The Indirect Evolutionary Game

5.4.1 Description of the Model and the Solution Concept

Consider the ultimatum game with subsequent production as described above.[9] Let X_1, respectively X_2 denote the pure strategy sets of the role–1–player, respectively the role–2–player. Accordingly, let t_i, $\Pi_i(.)$ and $U_i(.)$ for $i = 1, 2$ denote preference type, profit function and utility function of the role–i–player as defined above. The components X_i and $U_i(.)$, $i = 1, 2$ define a game $G^B(t_1, t_2)$ in which the role–1–player has type t_1 and the role–2–player has type t_2. Formally $G^B(t_1, t_2)$ is defined as:

$$G^B(t_1, t_2) \equiv \left((X_i, U_i(.))_{i=1,2}\right)\Big|_{(t_1,t_2)}$$

It will be referred to as the base game (given t_1, t_2) of the evolutionary game Γ, which will be defined in the following paragraph.

Suppose there is an infinitely large population of individuals who live for a single period. Each period all individuals of the respective generation are randomly matched in pairs to play the ultimatum game with subsequent production. According to the rules of $G^B(t_1, t_2)$ their strategy choices determine the utility they receive. After playing $G^B(t_1, t_2)$ individuals die and bequeath their preference type to the next generation. Evolutionary success is determined according to a function referred to as evolutionary success (or fitness) function. Specifically, it will be assumed that evolutionary success is equal to monetary payoff ($\Pi_1(.)$ respectively $\Pi_2(.)$). Formally, the indirect evolutionary game Γ is described by the following 8–tuple:

$$\Gamma \equiv ((X_i, T_i, U_i(.), \Pi_i(.))_{i=1,2})$$

with all components as described above. So, Γ consists of the base game $G^B(t_1, t_2)$, a description of the type spaces T_1 and T_2 as well the evolutionary success functions $\Pi_1(.)$ and $\Pi_2(.)$. T_1 and T_2 represent the mutation spaces for individual preferences of the role–1–player, respectively the role–2–player. Throughout it will be assumed that $T_1 = T_2 = \left[\frac{1}{2}, 1\right]$. The evolution of preference types t_i, which will be studied here, depends on inheritance and the occurrence of mutations.

In general, evolutionary success of an individual and its utility differ. It is a specific purpose of the indirect evolutionary approach to

[9] A more general description of symmetric as well as asymmetric indirect evolutionary games can be found in Königstein and Müller (forthcoming).

study evolution of preferences and behavior under such circumstances (see the discussion in the next section).

As a next step it will be shown in general terms how Γ can be analyzed and what is to be considered a solution of Γ. The first step in solving the indirect evolutionary game Γ is to solve $G^B(t_1, t_2)$ for all $(t_1, t_2) \in \left[\frac{1}{2}, 1\right]^2$. It is assumed that individuals choose strategies in order to maximize utility. Specifically, they play the subgame perfect equilibrium of $G^B(t_1, t_2)$. Notice that this part of the analysis was already done above. Let $\sigma = (s_1, q_1, \delta_2, q_2)$ represent a strategy profile in the game $G^B(t_1, t_2)$. It was shown that the SPE $(s_1^*, q_1^*, \delta_2^*, q_2^*)$ solely depends on technology parameters (c_1, c_2) and preference types (t_1, t_2). Since the evolution of types is to be studied here, let $\sigma^*(t_1, t_2)$ represent the subgame perfect equilibrium as a function of types for a given technology

$$\sigma^*(t_1, t_2) \equiv (s_1^*, q_1^*, \delta_2^*, q_2^*)|_{(c_1, c_2)} \, .$$

By evaluating $\Pi_i(\sigma)$ according to equilibrium play $\sigma^*(t_1, t_2)$ one gets the indirect evolutionary success function $\Pi_i^*(t_1, t_2) \equiv \Pi_i(\sigma^*(t_1, t_2))$. Together the components Π_i^* and T_i define a game

$$G^E \equiv ((T_i, \Pi_i^*)_{i=1,2})$$

which is the "direct evolutionary game" associated with the indirect evolutionary game Γ. G^E consists of a description of the type spaces and the indirect evolutionary success functions for each player. Notice that $\Pi_i^*(t_1, t_2)$ solely depends on the players' types. Importantly, G^E has all the ingredients of a usual asymmetric evolutionary game as described e.g. in Vega–Redondo (1996) and Weilbull (1995). Such games will be referred to as direct evolutionary games in order to distinguish them from indirect evolutionary games. After deriving G^E the analysis of Γ proceeds in the same manner as usual analyses of direct evolutionary games. Specifically a solution of G^E and thus a solution of Γ will be determined according to the notion of an evolutionarily stable strategy (ESS).

Notice that G^E is asymmetric. Since an individual k may be assigned either to role 1 or 2, it is important to distinguish between roles and individuals. In principle, individual k must be endowed with a preference type t_i^k for both roles i; i.e. k should be characterized by a type vector (t_1^k, t_2^k). However, although the formal description of Γ allows for role–dependent types, it will be assumed throughout that

$t_1^k = t_2^k$; i.e., an individual's degree of altruism does not depend on its role. So, every individual is completely characterized by a one dimensional parameter $t^k \in \left[\frac{1}{2}, 1\right]$. Thus a pairing $(t^k \times t^l) \equiv \left\{t^k, t^l\right\}$ of two individuals k and l may result either in match (t^k, t^l) — i.e. the t^k–type is player 1 in the ultimatum game — or in match (t^l, t^k) — i.e. the t^k–type is player 2 in the ultimatum game. Accordingly one has to distinguish $\sigma^*(t^k, t^l)$ and $\sigma^*(t^l, t^k)$ as well as $\Pi_i^*(t^k, t^l)$ and $\Pi_i^*(t^l, t^k)$ for $i = 1, 2$. It is customary in the analysis of asymmetric (direct) evolutionary games to determine the evolutionary success $\Pi^*(t^k \mid t^k \times t^l)$ of individual k in pairing $(t^k \times t^l)$ as the expected evolutionary success of both matches:

$$\Pi^*(t^k \mid t^k \times t^l) \equiv \frac{1}{2}\Pi_1^*(t^k, t^l) + \frac{1}{2}\Pi_2^*(t^l, t^k).$$

By this means the evolutionary success does not depend anymore on the role assignment. The game has been symmetrized. The notation will be simplified further. Namely, note that the evolutionary analysis is not concerned with the survival of individuals (k respectively l) but with preference types. Therefore, from now on $\Pi^*(\alpha \mid \alpha \times \beta)$ will denote the evolutionary success of preference type α in pairing $(\alpha \times \beta)$,i.e.

$$\Pi^*(\alpha \mid \alpha \times \beta) \equiv \frac{1}{2}\Pi_1^*(\alpha, \beta) + \frac{1}{2}\Pi_2^*(\beta, \alpha) \tag{5.6}$$

with $\alpha, \beta \in \left[\frac{1}{2}, 1\right]$. Accordingly $\Pi^*(\alpha \mid \alpha \times \alpha)$ is the evolutionary success of preference type α in pairing $(\alpha \times \alpha)$. Notice that from here on α (respectively β) does no longer identify role 1 (repectively role 2) as it was the case above in solving the base game.

Even though individuals can not choose their preference type, the evolutionary result will be determined as if it could be chosen as a strategy in the evolutionary game. So, within the evolutionary analysis preference types will be referred to as strategies.

Definition 5.2 *The parameter α is an evolutionarily stable strategy (ESS),*[10] *iff*

$$\Pi(\alpha \mid \alpha \times \alpha) \geq \Pi(\beta \mid \alpha \times \beta) \quad \forall \beta \in \left[\frac{1}{2}, 1\right] \tag{5.7}$$

[10] See e.g. Maynard Smith (1982), the survey by Hammerstein and Selten (1994), or the textbooks by Vega–Redondo (1996) and Weibull (1995).

$$\text{and } \forall \beta \neq \alpha \text{ with } \Pi(\alpha \mid \alpha \times \alpha) = \Pi(\beta \mid \alpha \times \beta) :$$
$$\Pi(\alpha \mid \alpha \times \beta) > \Pi(\beta \mid \beta \times \beta). \tag{5.8}$$

If the strict inequality in 5.8 is substituted by a weak inequality the conditions define a neutrally evolutionarily stable strategy (NESS).

In some cases when a strategy does not satisfy the conditions for ESS, it will be checked whether it does satisfy the weaker conditions of the NESS–concept. Thus, a preference type that exhibits the ESS or NESS property together with the induced base game strategy profile is considered a solution of the indirect evolutionary game Γ.

ESS and NESS are static solution concepts for indirect evolutionary games. In fact they treat the dynamics of the process of evolution only implicitely. Namely, the basic idea underlying evolutionary stability is that a population of individuals who are endowed with α should be stable against single mutants entering the population. Thus, if a mutant type β occurs it should not spread over time, which leads to the notion of ESS. For an example of a dynamic solution concept applied to an indirect evolutionary game see Königstein and Müller (forthcoming).

The solution of Γ will depend on the technology parameters c_1 and c_2. They will be assumed exogenous throughout. One could derive the evolutionary result for specific technologies. However, it will be a major issue in the analysis below to derive the evolutionary result for generic subsets of the technology space.

5.4.2 Some Methodological Remarks

A few comments on studying the evolution of preferences and the indirect evolutionary approach seem in place. Note that after σ^* has been determined as a mapping of α and β the indirect evolutionary approach proceeds in the same manner as the standard approach. Each player can be thought as being programmed to his type α, repectively β. A player's type represents no simple action or strategy, but a complex behavioral rule which depends also on the opponent player's type.

One might wonder about the dynamic process that characterizes social evolution in reality.[11] Sexual inheritance — i.e., taking evolutionary fitness as the number of children each type has — is a simple

[11] Binmore and Samuelson (1994, p. 48) suggest some care with regard to evolutionary dynamics. They remark that "stories about the replication of genes cannot be taken over without alteration".

story to illustrate how preferences are copied form one generation to the next. But, its dependence on economic profit is rather weak, at least nowadays. Also, this process would be very slow.[12]

It is probably more interesting to think of it as cultural evolution, e.g. via imitation of successful behavior. But, while one can easily imagine that agents imitate simple actions, it is hard to think of them copying preferences.[13]

A story that seems most plausible to me is one of norm formation. Suppose the society can be described by an overlapping generations model. Each period the same number of children are born with indefinite norms. Agents have to choose economic decisions only when they are old. The young generation observes the old generation's decisions, the norms which induce these decisions and their economic success. One can think of internalized social norms the same way as of individual preferences. The young adopt one of the norms when they grow old and norm adoption is positively related to economic success. Thus, instead of describing the analysis in terms of an evolutionary model one can think of it in terms of social dynamics.[14] A realistic parallel to this dynamic story may be seen in the development of a certain business ethics or corporate identity, which is adopted not only by employees within a specific firm, but which also spreads across firms.[15] However, the dynamic process is unmodelled in the game considered here and is left for further work.

5.5 Solution of the Indirect Evolutionary Game

5.5.1 Preliminaries

Unless specified differently, $(c_1, c_2) \in [0,1)^2$ will be assumed for the rest of the analysis. Other technologies, with $c_i \geq 1$ for one or both players are trivial and not very interesting, since they result in equal play by all types and no evolutionary pressure.

[12] See Selten (1991), p. 8.

[13] For a conceptual discussion of learning and evolution see Selten (1991).

[14] Binmore and Samuelson (1994, p. 48) give a different example for norm formation, namely that "in school and at home, we are actively brainwashed to conform to whatever social norms happen to be fashionable."

[15] Binmore, Swierzbinski, Hsu and Proulx (1993) even describe the dynamics of behavior within a laboratory experiment in terms of evolution.

Evolutionary success depends on $\sigma^*(\alpha,\beta)$ and $\sigma^*(\beta,\alpha)$; i.e., it depends on the subgame perfect equilibria induced by the players' preference types in both matches. Let

$$(\sigma^*_{kl},\sigma^*_{mn}) = [\sigma^*_{kl}(\alpha,\beta),\sigma^*_{mn}(\beta,\alpha)] \ , \text{ with } k,l,m,n \in \{0,1\}$$

denote a combination of SPE such that

$$q^*_1(\alpha,\beta) = k \ , \ q^*_2(\alpha,\beta) = l \ , \ q^*_1(\beta,\alpha) = m \ , \ q^*_2(\beta,\alpha) = n$$

It will be referred to as an SPE production vector. Thus e.g. $(\sigma^*_{10},\sigma^*_{11})$ says that pairing $(\alpha\times\beta)$ results in $(q^*_1 = 1, q^*_2 = 0)$ within match (α,β) and $(q^*_1 = 1, q^*_2 = 1)$ within match (β,α). It is a short hand for the production choices implied by a given equilibrium.

A match results either in $\sigma^*_{10},\sigma^*_{01}$ or σ^*_{11}. Therefore a pairing results in one of 9 possible SPE production vectors. While different pairings can result in the same production choices, their evolutionary payoffs differ, in general, since the equilibrium shares s^*_1 depend on the respective worker's type.

Lemma 5.1 *Within the support of any given SPE production vector, evolutionary success is non–decreasing in a players own egoism — i.e.* $\Pi(\alpha \mid \alpha\times\beta)$ *is non-decreasing in* α *for all* $(\alpha\times\beta)$ *that imply* $(\sigma^*_{kl},\sigma^*_{mn})$.

Proof. Recall,

$$\begin{aligned}
\Pi(\alpha \mid \alpha\times\beta) &= \frac{1}{2}\left[\Pi_1(\alpha,\beta) + \Pi_2(\beta,\alpha)\right] \\
\Pi_1(\alpha,\beta) &= s^*_1(\alpha,\beta)\cdot R\left[q^*_1(\alpha,\beta),q^*_2(\alpha,\beta)\right] - c_1\left[q^*_1(\alpha,\beta)\right] \\
\Pi_2(\beta,\alpha) &= (1 - s^*_1(\beta,\alpha))\cdot R\left(q^*_1(\beta,\alpha),q^*_2(\beta,\alpha)\right) - c_1\left(q^*_1(\beta,\alpha)\right)
\end{aligned}$$

By assumption, the production choices are unaffected such that:

$$\begin{aligned}
\frac{\partial \Pi_1(\alpha,\beta)}{\partial\alpha} &= \frac{\partial s^*_1(\alpha,\beta)}{\partial\alpha}\cdot R^* = 0, \\
\frac{\partial \Pi_2(\beta,\alpha)}{\partial\alpha} &= -\frac{\partial s^*_1(\beta,\alpha)}{\partial\alpha}\cdot R^* = \begin{cases} 0 & , \text{ if } s^*_1 = 1 \\ \frac{1-c_2}{(2\alpha-1)^2}\cdot R^* & , \text{ if } s^*_1 = \bar{s}_1 \end{cases}
\end{aligned}$$

and thus

$$\frac{\partial \Pi(\alpha \mid \alpha\times\beta)}{\partial\alpha} \geq 0.$$

More specifically, evolutionary success is strictly increasing in α within the support of any SPE production vector with $q_2^*(\beta,\alpha)=1$. Lemma 5.1 is important for the evolutionary analysis, since it implies that more egoistic mutations which enter a population without changing the production choices are at least as successful as the incumbant. Such mutations do strictly better within the support of any SPE production vector with $q_2^*=1$ ($\Rightarrow s_1^*=\bar{s}_1$) in one or both matches.

So, there is some evolutionary pressure toward egoism. It arises from the fact that more egoistic workers get higher shares in the bargaining part of the game. I.e., the evolutionary advantage of egoistic players is due to extracting more from a given surplus than altruistic types. The game theoretic analysis, however, showed that less egoism may increase the surplus that gets distributed. This indicates, why such behavior might also be successful in the evolutionary game. Namely, because the efficiency gains it generates may offset the smaller shares from bargaining.

The analysis proceeds by identifying a set of types which are candidates for an ESS. Note that every ESS must be a best reply to itself, such that only these types are ESS–candidates. Symmetric pairing results in symmetric play — i.e. equal play in both matches. There are three symmetric SPE production vectors: $(\sigma_{11}^*,\sigma_{11}^*)$, $(\sigma_{01}^*,\sigma_{01}^*)$ and $(\sigma_{10}^*,\sigma_{10}^*)$. Thus, the set of candidates will be determined for each of the three cases.

Lemma 5.2 *Given* $(c_1,c_2)\in[0,1)^2$, *if there exists an ESS* α *sustaining* $(\sigma_{11}^*,\sigma_{11}^*)$, *it is one of the following candidates:*

$$\alpha^0=1$$
$$\alpha^1=\frac{2-c_2}{1+c_2}$$
$$\alpha^2=\frac{1}{c_1+c_2}$$

(where α^1 *implies* $\Delta U_1^2(\alpha,\alpha)=0$ *and* α^2 *implies* $\Delta S_1(\alpha,\alpha)=0$*).*

Proof. Lemma 5.1 implies that α must be maximal in the support of $(\sigma_{11}^*,\sigma_{11}^*)$. The support of $(\sigma_{11}^*,\sigma_{11}^*)$ hinges on $\Delta U_1^2\leq 0$ and $\Delta S_1\leq 0$. Since $\Delta U_1^2(\alpha,\alpha)$ and $\Delta S_1(\alpha,\alpha)$ are increasing in α an ESS must be exactly on one or both of the boundaries ($\Rightarrow \alpha^1$ or α^2), unless $\Delta U_1^2(\alpha,\alpha)<0$ and $\Delta S_1(\alpha,\alpha)<0$ for all α. In the latter case α^0 sustains full efficiency and is the candidate for an ESS.

So, for a given technology there exists at most one type who reaches full efficiency and may be an ESS. Furthermore, it is one of the three candidates α^0, α^1 and α^2. By a similar reasoning one can determine the candidates sustaining $(\sigma^*_{01}, \sigma^*_{01})$. This requires $\Delta S_1(\alpha, \alpha) > 0$ and $\Delta U^1_{\bar{1}}(\alpha, \alpha) \leq 0$ such that one candidate is

$$\alpha^3 = \frac{1 - c_2}{1 - c_1}$$

For α^3 the condition for $\Delta U^1_{\bar{1}}$ is binding — i.e. $\Delta U^1_{\bar{1}}(\alpha^3, \alpha^3) = 0$. If $\Delta S_1(\alpha, \alpha) > 0$ and $\Delta U^1_{\bar{1}}(\alpha, \alpha) > 0$ for all α the candidate sustaining $(\sigma^*_{01}, \sigma^*_{01})$ is α^0 again.

Regarding $(\sigma^*_{\bar{1}0}, \sigma^*_{\bar{1}0})$ the following proposition holds.

Proposition 5.4 *Generically, there exists no ESS supporting* $(\sigma^*_{\bar{1}0}, \sigma^*_{\bar{1}0})$ *— i.e. it may exist only for technologies which have zero measure relative to the unit square.*

Proof. Suppose otherwise and let α be an ESS supporting $(\sigma^*_{\bar{1}0}, \sigma^*_{\bar{1}0})$. The proof of lemma 5.1 shows that within the support of $(\sigma^*_{\bar{1}0}, \sigma^*_{\bar{1}0})$ more egoistic mutations earn the same as an incumbant. The reason is that in this case $s^*_{\bar{1}} = 1$ in both matches. Thus, the demanded share and therefore the evolutionary payoff is not varying in type.

Note further that $\Delta S_1(\alpha, \alpha)$, $\Delta U^1_{\bar{1}}(\alpha, \alpha)$ and $\Delta U^2_{\bar{1}}(\alpha, \alpha)$ are monotonically increasing in α. Thus, except for degenerate technologies, if there is any type α at all inducing $(\sigma^*_{\bar{1}0}, \sigma^*_{\bar{1}0})$ when paired with itself, then there exist another type β in the neighborhood of α which induces $(\sigma^*_{\bar{1}0}, \sigma^*_{\bar{1}0})$ when paired with itself as well. Thus, a mutation β earns the same evolutionary payoff as an incumbant α:

$$\Pi(\beta \mid \alpha \times \beta) = \Pi(\alpha \mid \alpha \times \alpha). \tag{5.9}$$

Furthermore, a mutation α earns the same evolutionary payoff as an incumbant β:

$$\Pi(\alpha \mid \alpha \times \beta) = \Pi(\beta \mid \beta \times \beta). \tag{5.10}$$

Together 5.9 and 5.10 contradict the claim that α is an ESS (see 5.8).

Within the support of $(\sigma^*_{\bar{1}0}, \sigma^*_{\bar{1}0})$ one may therefore look for types which are NESS instead of ESS. Following the arguments above there exist (generically) at most two evolutionarily stable preference types for a given technology (c_1, c_2): one supporting $(\sigma^*_{\bar{1}1}, \sigma^*_{\bar{1}1})$ and one supporting $(\sigma^*_{01}, \sigma^*_{01})$.

5.5.2 Existence of Evolutionarily Stable Altruistic Preferences

Now, the goal of the analysis shall be stated more specifically: Within this subsection the main result of the paper will be established. It is a theorem claiming that there exist generic technologies such that only altruistic preferences are evolutionarily stable. Within the next section it will be investigated for all generic technologies within the unit square:

- whether $\alpha < 1$ in the support of $(\sigma^*_{11}, \sigma^*_{11})$ — i.e. altruistic preferences that sustain full efficiency — can be evolutionarily stable and
- whether or not α^0 — i.e. egoism — is an ESS or NESS.

The procedure is as follows:

1. *Candidates:* For a given subset of the technology space the candidate for an ESS supporting $(\sigma^*_{11}, \sigma^*_{11})$ will be determined. Also, play $\sigma^*_{kl}(\alpha^0, \alpha^0)$ will be determined. Since $\alpha > \alpha^0$ is not feasible, α^0 is necessarily maximal in the support of $(\sigma^*_{kl}, \sigma^*_{kl})$ and is therefore a candidate for an ESS or NESS in any case.
2. *Feasible Play:* For each candidate it will be investigated which plays are feasible via mutation and which mutation is maximal in the support of each feasible play. Such mutations will be called "challengers". This step of the procedure has to be done separately for each ot the two matches (α, β) and (β, α).
3. *Success of Entry:* It will be checked whether any mutation is improving — i.e. whether it earns an evolutionary payoff that is at least as high as that of the ESS– or NESS candidate. According to lemma 5.1 it suffices to do this only for the challengers.

Thus the analysis will be done for separate subsets of the technology space. It will be partitioned according to the following conditions:

$$c_1 + c_2 \leq 1 \Longrightarrow \Delta S_1(\alpha, \alpha) \leq 0 \forall \alpha \in \left[\frac{1}{2}, 1\right] \tag{5.11}$$

$$c_2 \leq c_1 \Longrightarrow \Delta U_1^1(\alpha, \alpha) \leq 0 \forall \alpha \in \left[\frac{1}{2}, 1\right] \tag{5.12}$$

$$c_2 \leq \frac{1}{2} \Longrightarrow \Delta U_1^2(\alpha, \alpha) \leq 0 \forall \alpha \in \left[\frac{1}{2}, 1\right] \tag{5.13}$$

$$c_1 \leq \frac{1 + c_2^2 - c_2}{2 - c_2} \Longrightarrow \alpha^1 \leq \alpha^2 \tag{5.14}$$

Table 5.2. Subsets of the Technology Space

Technology Set	Restrictions		
$T_{1.a}$	$c_1 + c_2 \leq 1,$	$0 \leq c_2 \leq \frac{1}{2}$	
$T_{1.b}$	$c_1 + c_2 \leq 1,$	$0 \leq c_1 \leq \frac{1}{2} < c_2 < 1$	
$T_{2.a}$	$c_1 + c_2 > 1,$	$0 \leq c_2 \leq \frac{1}{2} \leq c_1 < 1$	
$T_{2.b}$	$c_1 + c_2 > 1,$	$\frac{1}{2} \leq c_2 \leq c_1 < 1$	
$T_{2.c.1}$	$c_1 + c_2 > 1,$	$\frac{1}{2} \leq c_1 < c_2 < 1,$	$c_1 \geq \frac{1+c_2^2-c_2}{2-c_2}$
$T_{2.c.2}$	$c_1 + c_2 > 1,$	$\frac{1}{2} \leq c_1 < c_2 < 1,$	$c_1 \leq \frac{1+c_2^2-c_2}{2-c_2}$
$T_{2.d}$	$c_1 + c_2 > 1,$	$0 \leq c_1 \leq \frac{1}{2} < c_2 < 1$	

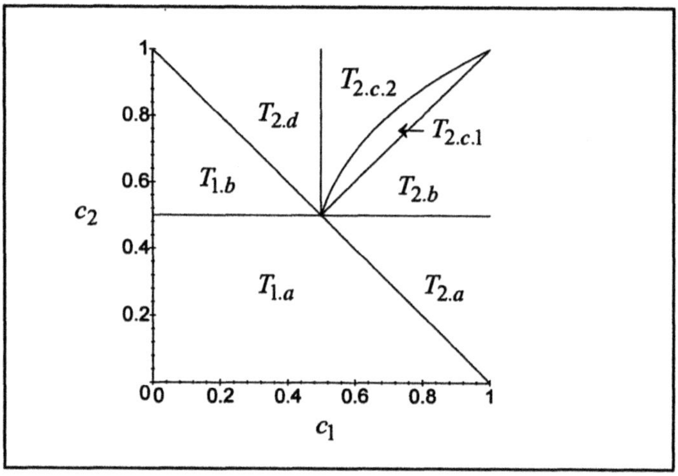

Fig. 5.4. Subsets of the Technology–Space

If 5.11 holds this can be interpreted as situation where joint effort of firm owner and worker is highly productive. Condition 5.12 says that in this case the firm owner's productivity $(1 - c_1)$ is less than that of the worker and technologies satisfying 5.13 are such that the worker is highly productive, i.e., $(1 - c_2) \geq \frac{1}{2}$. Condition 5.14 is a technical condition that has no deeper interpretation. It just separates technologies where α^1 may be the ESS-candidate supporting efficient play from those where α^2 may be this candidate. The subsets of the technology space shall be defined as shown in table 5.2: The sets are represented by different regions in figure 5.4.

When determining whether a candidate sustaining efficiency is an ESS the following lemma is useful.

Lemma 5.3 *Suppose α is an ESS candidate sustaining $(\sigma^*_{\bar{1}1}, \sigma^*_{\bar{1}1})$. Then, no mutation $\beta < \alpha$ can be improving: $\Pi(\beta \mid \alpha \times \beta) \leq \Pi(\alpha \mid \alpha \times \alpha)$.*

The proof is given in the appendix. Lemma 5.3 implies that to check whether candidate α, with $(\alpha \times \alpha) \Rightarrow (\sigma^*_{\bar{1}1}, \sigma^*_{\bar{1}1})$, is an ESS it suffices to investigate mutations $\beta > \alpha$.

Furthermore, notice that extreme altruism can not be an ESS or a challenger mutation. Namely, the game theoretic solution implies that in match (α, β) neither $\alpha < \frac{1}{1+c_1}$ nor $\beta < \frac{1}{1+c_2}$ can be maximal in the support of $\sigma^*(\alpha, \beta)$. Since $c_i < 1$ is assumed, it follows that both, α and β, are bounded away from $\frac{1}{2}$ and therefore $\alpha > \frac{1}{2}$ and $\beta > \frac{1}{2}$ can be assumed throughout the analyis.

Proposition 5.5 *If egoism allows for efficient production, that is, if $(\alpha^0 \times \alpha^0)$ implies $(\sigma^*_{\bar{1}1}, \sigma^*_{\bar{1}1})$, then only egoistic preferences are evolutionarily stable.*

Proof. Since $\Delta S_1(\alpha^0, \alpha^0) = c_1 + c_2 - 1$ and $\Delta U^2_{\bar{1}}(\alpha^0, \alpha^0) = 2c_2 - 1$, it follows: an α^0-population sustains $(\sigma^*_{\bar{1}1}, \sigma^*_{\bar{1}1})$, if and only if $(c_1, c_2) \in T_{1.a}$. Conditions 5.11 and 5.13 imply $\Delta S_1(\alpha, \alpha) \leq 0$ and $\Delta U^2_{\bar{1}}(\alpha, \alpha) \leq 0$ $\forall \alpha \in \left[\frac{1}{2}, 1\right]$ such that α^0 is the only ESS-candidate (lemma 5.1).

Lemma 5.3 implies that mutations $\beta < \alpha$ earn strictly less and since $\beta > 1$ is impossible, α^0 is the unique ESS.

The intuition for proposition 5.5 is quite simple. In principle, altruistic preferences might be superior to egoistic preferences by increasing productive efficiency. But, if egoism reaches maximal efficiency, altruism has no leverage. It's important to notice, however: even though all types reach full productive efficiency in case $(c_1, c_2) \in T_{1.a}$ only the egoistic type is an ESS.

Proposition 5.5 characterizes the solution of the evolutionary game for all $(c_1, c_2) \in T_{1.a}$. Within this subset of the technology space altruism can not survive in the long run. But, there are other subsets where only altruistic preferences may be evolutionarily stable. This is the content of the following theorem.

Theorem 5.1 *In the indirect evolutionary game as described here, there exist generic subsets within the unit square of the technology space such that only altruistic preferences — i.e. $\alpha < 1$ — are evolutionarily stable.*

Proof. Suppose $(c_1, c_2) \in T_{1.b}$. The set $T_{1.b}$ has positive measure relative to the technology space and it will be shown that $(c_1, c_2) \in T_{1.b}$ implies:

$$\alpha^1 = \tfrac{2-c_2}{1+c_2} < 1 \text{ sustains } (\sigma^*_{11}, \sigma^*_{11}) \text{ and is an ESS and} \tag{5.15}$$

$$\alpha^0 \text{ is no ESS.} \tag{5.16}$$

To see this, note first that 5.11 implies $\Delta S_1(\alpha, \alpha) \leq 0 \;\forall \alpha \in \left[\frac{1}{2}, 1\right]$. Therefore α^1 and α^0 are the only ESS-candidates. While α^1 supports efficient play $(\sigma^*_{11}, \sigma^*_{11})$, α^0 supports $(\sigma^*_{10}, \sigma^*_{10})$.

Candidate: α^1

To check whether α^1 is an ESS it is sufficient to investigate mutations $\beta > \alpha^1$ (lemma 5.3).

Feasible Plays, Match (α, β): Since $\Delta S_1(\alpha^0, \alpha^0) \leq 0$ and $\Delta S_1(\alpha, \beta)$ is increasing in α and β:

$$\Delta S_1(\alpha^1, \beta) \leq 0 \forall \beta \in \left[\alpha^1, 1\right]. \tag{5.17}$$

Since $\Delta U^2_1(\alpha^1, \alpha^1) = 0$ and $\Delta U^2_1(\alpha, \beta)$ strictly increasing in β:

$$\Delta U^2_1(\alpha^1, \beta) > 0 \forall \beta \in \left(\alpha^1, 1\right]. \tag{5.18}$$

5.17 and 5.18 imply that all mutations $\beta > \alpha^1$ lead to play $\sigma^*_{10}(\alpha^1, \beta)$.

Feasible Plays, Match (β, α): Since $\Delta S_1(\alpha^0, \alpha^0) \leq 0$ and $\Delta S_1(\beta, \alpha)$ increasing in α and β:

$$\Delta S_1(\beta, \alpha^1) \leq 0 \forall \beta \in \left[\alpha^1, 1\right]. \tag{5.19}$$

Since $\Delta U^2_1(\alpha^1, \alpha^1) = 0$ and $\Delta U^2_1(\beta, \alpha^1)$ strictly increasing in β given $c_2 < 1$:

$$\Delta U^2_1(\beta, \alpha^1) > 0 \forall \beta \in \left(\alpha^1, 1\right]. \tag{5.20}$$

5.19 and 5.20 imply that all mutations $\beta > \alpha^1$ induce play $\sigma^*_{10}(\beta, \alpha^1)$.

So, all mutations $\beta > \alpha^1$ result in SPE production vector $(\sigma^*_{10}, \sigma^*_{10})$. Lemma 5.4 implies that all these mutations earn the same. Since

$$\Pi(\beta \mid \alpha^1 \times \beta) = \frac{1}{2}(1 - c_1) \text{ and}$$
$$\Pi(\alpha^1 \mid \alpha^1 \times \alpha^1) = \frac{1}{2}(2 - c_1 - c_2)$$

it follows

$$\Pi(\beta \mid \alpha^1 \times \beta) < \Pi(\alpha^1 \mid \alpha^1 \times \alpha^1) \quad \forall \beta > \alpha^1.$$

This proves 5.15.

Candidate: α^0

To show that α^0 is no ESS in case of $T_{1.b}$ it suffices to find an example for an improving mutation. Suppose a mutation $\beta = \frac{1}{2c_2}$. One can easily verify that in this case $\Delta S_1(\alpha^0, \beta) \leq 0$, $\Delta U_1^2(\alpha^0, \beta) \leq 0$, $\Delta S_1(\beta, \alpha^0) \leq 0$ and $\Delta U_1^2(\beta, \alpha^0) > 0$. Accordingly, $(\alpha^0 \times \beta)$ implies SPE production vector $(\sigma_{11}^*(\alpha^0, \beta), \sigma_{10}^*(\beta, \alpha^0))$, while $(\alpha^0 \times \alpha^0)$ resulted in $(\sigma_{10}^*, \sigma_{10}^*)$. Hence:

$$\Pi(\beta \mid \alpha^0 \times \beta) = \frac{1}{2}\left[(1 - c_1) + 2\left(1 - s_1^*(\alpha^0, \beta)\right) - c_2\right],$$

$$\Pi(\alpha^0 \mid \alpha^0 \times \alpha^0) = \frac{1}{2}(1 - c_1)$$

such that

$$\Pi(\beta \mid \alpha^0 \times \beta) > \Pi(\alpha^0 \mid \alpha^0 \times \alpha^0) \Longleftrightarrow 1 - c_2 > 0 \tag{5.21}$$

which holds for all $(c_1, c_2) \in T_{1.b}$.

The theorem establishes that altruism can drive out egoism via evolutionary forces. The reason is that social concern increases productive efficiency. More specifically, if an egoist, $\beta = 1$, enters an α^1–population, he induces play $(\sigma_{10}^*, \sigma_{10}^*)$, while the incumbants play $(\sigma_{11}^*, \sigma_{11}^*)$ when paired with each other. Thus, within pairing $(\beta \times \alpha^1)$ both, entrant and incumbant, earn the same evolutionary payoff, but incumbants earn more when paired with one another such that their average payoff is higher than that of the entrant.

The intuition for the non–stability of egoistic preferences is slightly more complicated. If a sufficiently altruistic type, e.g. $\beta = \frac{1}{2c_2}$, enters an egoistic population he induces σ_{10}^* in match (β, α^0), but σ_{11}^* in match (α^0, β). Thus, if the altruistic type has the role of a firm owner, he treats the egoistic worker the same way as an egoistic firm owner does. But, if the altruistic type has the role of a worker, full efficiency is reached. An altruistic worker needs a smaller return share than an egoist to supply full effort. The egoistic firm owner grants this share, since by doing so he is better off as well.[16] Thus the worker's altruism is profitable for both, the worker and the egoistic firm owner. Again, the key point for evolutionary stability is that the altruistic type's average payoff is higher than the incumbant's which is captured in condition 5.21.

Theorem 5.1 says that non–egoistic preferences can develop and be sustained within a society for a generic domain of the model. As shown below this does not only hold for $(c_1, c_2) \in T_{1.b}$, but is the case for larger subsets of the technology space.

5.5.3 Evolutionary Results for All Technologies within the Unit Square

The following propositions characterize, for which technological conditions within the unit square, efficient production can be sustained as evolutionarily stable and whether or not egoism is an ESS or a NESS. The proofs are contained in the appendix, where the results are worked out for each subset of the technology space.

Proposition 5.6 *Preference type $\alpha^1 = \frac{2-c_2}{1+c_2}$ sustains $(\sigma^*_{\tilde{1}1}, \sigma^*_{\tilde{1}1})$ and is an ESS, if the following conditions hold:*

$$c_2 > \frac{1}{2},$$
$$\alpha^1 \leq \alpha^2 \Longleftrightarrow c_1 \leq \frac{1 + c_2^2 - c_2}{2 - c_2},$$
$$c_1 < \frac{4}{3} - \frac{2}{3} c_2.$$

Proof. This follows from theorem 5.1, lemma D.4 and lemma D.5 (see the appendix).

[16] Actually, in the given example with $\beta = \frac{1}{2c_2}$ being the entrant, the incumbant α^0 is indifferent between $\sigma^*_{\tilde{1}1}(\alpha^0, \beta)$ and $\sigma^*_{\tilde{1}0}(\alpha^0, \beta)$. Namely, $\beta = \frac{1}{2c_2}$ implies $\bar{s}_1 = \frac{1}{2}$ such that $\Pi_1(\alpha_0, \beta) = \frac{1}{2} - c_1$ in both types of play. But this is just due to $\beta = \frac{1}{2c_2}$ being a limiting case. Suppose a different mutation $\beta^- < \beta$ such that β^- is marginally smaller than β. Accordingly, $\bar{s}_1(\beta^-) = \frac{1}{2} + \varepsilon$ where ε is a positive number. Thus play $\sigma^*_{\tilde{1}1}(\alpha^0, \beta)$ yields the α^0–type a payoff $\Pi_1(\alpha_0, \beta) \mid \sigma^*_{\tilde{1}1} = \left(\frac{1}{2} + \varepsilon\right) 2 - c_1 = \frac{1}{2} - c_1 + 2\varepsilon > \frac{1}{2} - c_1$. Thus unless a limiting mutation — i.e. a challenger — enters, the incumbant's profit strictly increases with increased efficiency.

Proposition 5.7 *Preference type $\alpha^2 = \frac{1}{c_1+c_2}$ sustains $(\sigma^*_{11}, \sigma^*_{11})$ and is an ESS, if the following conditions hold:*

$$c_1 + c_2 > 1 \ ,$$
$$\frac{1 + c_2^2 - c_2}{2 - c_2} \leq c_1 \Longleftrightarrow \alpha^2 \leq \alpha^1 \text{ and}$$
$$1 - c_2 - (2 - c_1 - c_2)^2 < 0.$$

Proof. This follows from lemmata D.1, D.2 and D.3 (see the appendix).

Proposition 5.8 *Preference type $\alpha^0 = 1$:*
(1) sustains $(\sigma^*_{11}, \sigma^*_{11})$ and is an ESS, if and only if:

$$c_1 + c_2 \leq 1 \text{ and}$$
$$0 \leq c_2 \leq \frac{1}{2}.$$

(2) sustains $(\sigma^*_{01}, \sigma^*_{01})$ and is an ESS if:

$$c_2 \leq c_1 \text{ and}$$
$$1 - \frac{1}{2}c_2 < c_1.$$

(3) sustains $(\sigma^*_{10}, \sigma^*_{10})$ and is not a NESS if:

$$c_2 > \frac{1}{2} \ ,$$
$$c_2 > c_1 \text{ and}$$
$$2 - 2c_1 > c_2 \Longleftrightarrow 1 - \frac{1}{2}c_2 > c_1.$$

Proof. The results follow from propositions 5.4 and 5.5, theorem 5.1 as well as lemmata D.1, D.2, D.3, D.4 and D.5 (see the appendix).

5.5.4 Properties of the Evolutionary Results

The results are illustrated in figures 5.5 to 5.7.

Regions within the technology space for which α^1, respectively α^2, are evolutionary stable are displayed in figure 5.5. Proposition 5.8 is illustrated in figure 5.6. In cases were α^0 sustains $(\sigma^*_{10}, \sigma^*_{10})$ it was investigated under which conditions it is **not** neutrally evolutionarily stable, which is simpler than to prove whether α^0 is indeed a NESS. But, it is clear from proposition 5.4 that α^0 is no ESS in these cases

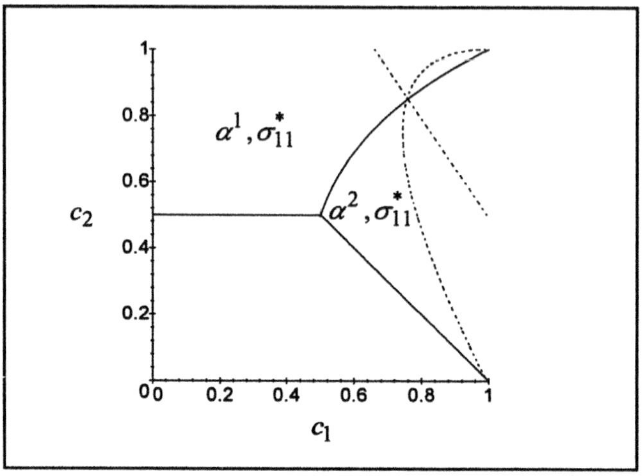

Fig. 5.5. ESS-Regions of Altruistic Preferences.

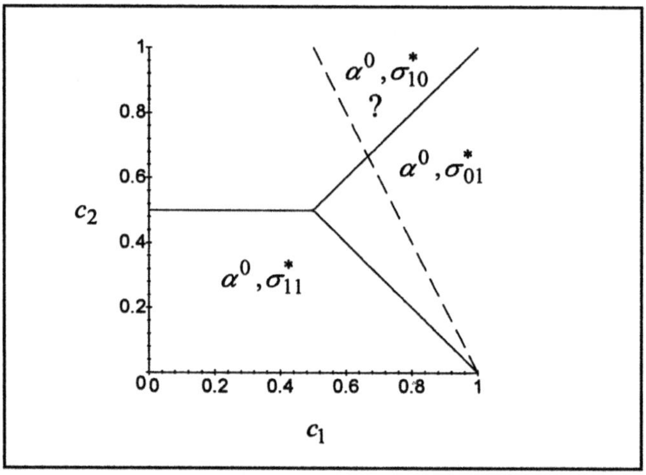

Fig. 5.6. Play Induced by Egoism, Corresponding ESS–Regions and Region were α^0 is a NESS, at best (Indicated by Question Mark).

anyway and that it is a NESS, at best. Proposition 5.8 reduces these cases to the region displaying the question mark.

Figures 5.5 and 5.6 are superimposed in figure 5.7. One can see that egoism is evolutionarily stable either if it reaches productive efficiency or if the firm owner's net productivity $(1 - c_1)$ is sufficiently low. More exactly, only egoistic preferences are ESS at these extremes. In between

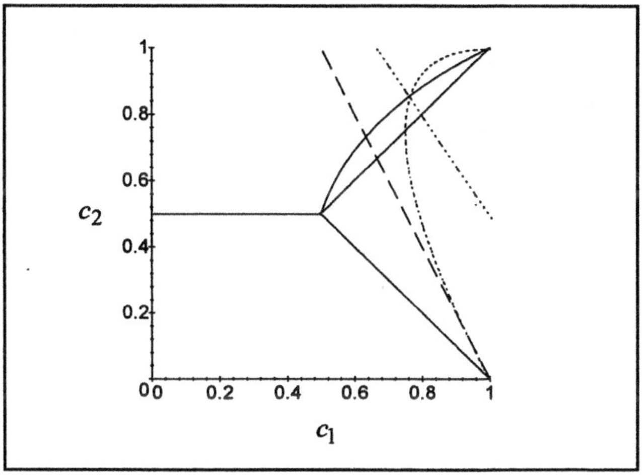

Fig. 5.7. Display of Both Previous Figures Together.

there is a region were altruism can evolve. It was not at all obvious from the start that such a region exists. I regard the identification of the three regions as one of the major qualitative result of this study.

The ESS regions are non–exclusive. There is some overlap, which means that for some technologies altruistic preferences as well as egoistic ones are evolutionary stable. Multiple evolutionarily stable strategies are not unusual in evolutionary games. E.g. in games with multiple strict Nash equilibria like in coordination games all respective equilibrium strategies are evolutionarily stable.[17] Proposition 5.4 has shown that there are at most two ESS for a given technology.

Regarding the uniqueness of the results, one should notice further that for some subsets of the technology space there is another candidate for an ESS which wasn't investigated. This is $\alpha^3 = \frac{1-c_2}{1-c_1}$, which is a candidate sustaining production vector $(\sigma^*_{01}, \sigma^*_{01})$. From theorems 5.1 it is clear that α^3 is no ESS given technology $T_{1.a}$. Also, $\Delta U_1^1(\alpha, \alpha) \leq 0$ can only be binding for $c_2 \geq c_1$ such that α^3 is a candidate only for technologies within $T_{1.b}$, $T_{2.c.1}$, $T_{2.c.2}$ and $T_{2.d}$. Finally, since an α^3–population does not reach full productive efficiency, one should expect that it can be invaded easily; either by egoistic types — which imply

[17] The ESS-concept just asks whether a given population is stable against a small number of entrants. It does not claim that this population evolves from arbitrary starting points of the evolutionary process. Thus, it is not surprising that more than one strategy exhibits the ESS-property.

inefficiency as well, but capture higher bargaining shares — or by more altruitic types — who reach full efficiency.

All evolutionarily stable strategies proposed here are strictly superior to every mutation — i.e., the ESS–inequality 5.7 holds strictly. While this supports the robustness of the results it was mainly done for convience. If the technology is such that the evolutionary payoffs of incumbant and entrant are equal, one would have to check ESS–condition 5.8. In essence, this means examining whether or not an ESS–region contains its boundary and therefore it was left out.

Both, α^1 and α^2, are continuous functions of the technology parameters. Both converge to the value 1, if the technology approaches $T_{1.a}$, and to the value $\frac{1}{2}$, if (c_1, c_2) converges to $(1, 1)$. While it was argued above that extreme altruism $\alpha = \frac{1}{2}$ can not be an ESS, α^1 can get arbitrarily close to $\frac{1}{2}$.

Since α^1 and α^2 are unique for a given technology, the ESS–support of a specific value $\bar{\alpha} < 1$ has zero measure in the unit square of the technology space, which might be viewed as a shortcoming. It is due to the specification of the model: namely, the evolutionary results are derived for a fixed technology. Alternatively, one could think of a more general model were both, technology and preferences, are allowed to change, e.g. cost might be stochastic. Given the framework of such an extended model, the analysis here is a special case, in which technological change is sufficiently slow such that evolution settles. One might wonder whether the main results of the analysis here still hold within a model with stochastic cost. — I think, they do! And I will give some intuition for this conjecture, leaving the details for future work.

Specifically, suppose the cost vector is randomly drawn from a probability distribution on a subset C of the technology space with C exhibiting positive measure relative to the unit square. I will argue, that there exist such subsets for which only $\bar{\alpha} < 1$ is an ESS. For instance, let C be a subset of $T_{1.b}$ with $\underline{c}_i \leq c_i \leq \bar{c}_i$ for $i = 1, 2$, and let (c_1, c_2) be randomly drawn from a continuous two–dimensional probability distribution on C. Furthermore, define $A = [\bar{\alpha}^1, \underline{\alpha}^1]$ with

$$\bar{\alpha}^1 = \frac{2 - \bar{c}_2}{1 + \bar{c}_2} \text{ and } \underline{\alpha}^1 = \frac{2 - \underline{c}_2}{1 + \underline{c}_2}.$$

Accordingly, $\alpha \in A$ implies $\alpha < 1$ and, due to proposition 5.6, each $\alpha \in A$ is an ESS in the deterministic model for some technology $(c_1, c_2) \in C$. Furthermore any $\alpha \notin A$ is not an ESS of the deterministic model. Thus, for generic probability distributions on C, an

ESS of the stochastic model will be in the interior of A. Following these loose arguments, altruism may also be evolutionarily stable in a world were technology changes more rapidly than the evolutionary process settles. But, once more, it remains to check the details.

5.6 Summary

The game theoretic analysis showed that agents with social preferences can achieve efficiency in production while egoistic agents may not - given the same production technology.

Within the evolutionary analysis the main question was, whether social preferences and prosocial behavior may be evolutionarily stable. The answer is affirmative. It was shown that this holds not only for special technologies, but for generic subsets of the relevant technology space.

Egoism is also an ESS, if it achieves full productive efficiency — as one should expect — and if the firm owner's net productivity is sufficiently low. In the latter case, if an altruistic type enters an egoistic population, the efficiency gains which he reaches as firm owner via joint production, are smaller than the payoff loss he incurs due to the reduced share he gets as a worker.

Overall, the formal analysis confirms the conjecture that social preferences may evolve and be sustained within societies, because they generate economic efficiency. The social production technology and the bargaining rules as well as the utility functions and the evolutionary success functions are rather simple, which I consider as a strength of the model. It captures some intuition and some stylized empirical facts, e.g. regarding heterogenous demands in bargaining.

The model does not allow for disagreements. Furthermore the preference types are common knowledge, which is a questionable assumption. Introducing incomplete information regarding the types would complicate the analyis, but as long as there is sufficient revelation of types — i.e., not necessarily perfect revelation — I conjecture that the qualitative results go through.

One might get disagreements as evolutionarily stable behavior by introducing a stochastic (but privately known) disagreement value $v_i \in V_i$ — i.e. an outside option value which player i receives in case of rejection $\delta_2 = 0$ and which is only known by i. In this case, player 1's equilibrium demand $s_1^* = \bar{s}_1$ — which is determined in the first part of the indirect evolutionary analysis — will maximize player 1's expected utility given his expectation regarding player 2's outside option value. One can easily imagine that s_1^* does not satisfy the participation constraint $U_2(\delta_2 = 0) \geq v_2$ for all outside option values v_2. Within the model, here, it was implicitely assumed that outside option values are zero, which simplified the analysis considerably. Investigating the influence of positive outside option values is left for future research.

How far one likes to interprete these results is to some extent a matter of taste. I personally view the model as archetypical of socio–economic conficts. It describes consistently how prosocial behavior can emerge and be sustained within a society. In this aspect — namely, in the object of study — it is similar to what Sugden (1993) refers to as "the contractarian enterprise", which inquires "how rational individuals, each concerned with his or her own interests, might agree to cooperate for mutual benefit".[18]

But it differs in a rather important aspect; i.e., the explanation of norm compliance. E.g. Hobbes (1651) and Buchanan (1975) rely on a government to enforce norm compliance of the agents. Alternatively, according to Gauthier (1986) an agent may voluntarily impose constraints on his actions, since this may maximize some long term interests. His line of arguments can be descibed as follows: Keeping a promise in return for a received favor is advantageous, since people's dispositions for norm compliance are not opaque to one another. Any signal of the willingness not to act reciprocally will preclude reaching an efficient agreement.[19]

Notice, firstly, that the revelation of agents' types assumed by Gauthier is similar to what is modelled here. But, secondly, in his case a player's type is not a specific individual characteric, but changes with his intended action. In the model presented here compliance is modelled via preferences which guarantee it. This is conceptually simple and seems intuitive to me. Furthermore, here, the agent does not have to trade–off the long term versus the short term in his mind; the long–term is taken care of by evolution. Finally, the evolutionary analysis investigates whether a society complying to a certain norm is robust against a small number of entrants who exhibit a different norm, which is surely appealing.

[18] Sugden (1993), p.1.

[19] This explanation for the existence of social norms is closely related to that of Axelrod (1986) who studies the evolution of norms with the use of computer simulations.

Contractarianism avoids to characterize what a good society is, and according to Sugden (1993) it can be regarded as a step away from the Kantian view on norms.[20] The indirect evolutionary analysis might be regarded even a further step away. Here, the behavior that persists in the long run does not follow the rationale of a government nor does it necessarily preserve the long term interests of individuals. It is just the behavior that is most successful according to the evolutionary criterion.

[20] See Sugden (1993), p. 4 and p. 9.

A. Experimental Instructions Chapter 1

This appendix contains the German instructions to the experiment described in chapter 1 as well as an English translation. Furthermore it includes the 16PA–postexperimental questionnaire (German version).

Allgemeine Hinweise

Für Ihre Teilnahme an diesem Experiment bekommen Sie 5,- DM geschenkt. In Abhängigkeit von Ihrem Erfolg können Sie zusätzliches Geld gewinnen, aber natürlich auch Geld verlieren. Ihre Gewinne werden Ihnen tatsächlich ausgezahlt. Verluste werden einkassiert bzw. mit Gewinnen aus anderen Spielen und Ihrer Teilnahmeprämie verrechnet. In allen Spielen haben Sie die Möglichkeit, so zu entscheiden, daß Sie jedes finanzielle Risiko vermeiden.

Zur Vereinfachung werden alle Zahlungen zunächst nur im Computer verbucht und später durch das Sekretariat von Herrn Professor Dr. Reinhard Tietz, Raum 110 C abgewickelt. Sie können sich Ihr Geld dort während des Semesters jeweils mittwochs, 10 bis 12 Uhr oder 13 bis 15 Uhr, abholen. Gewinne, welche bis Ende des WS 93/94 nicht abgeholt werden, werden nicht mehr ausgezahlt.

Ihre Teilnehmernummer lautet:

Sie benötigen dieses Hinweisblatt mit Ihrer Teilnehmernummer als Ausweis für die Geldauszahlung. Nehmen Sie es am Ende des Experiments mit nach Hause und bringen Sie es mit, wenn Sie Ihren Gewinn abholen.

Wir bitten Sie, während des Experimentes nicht miteinander zu reden. Wenn Sie etwas nicht verstehen oder sonstige Fragen stellen möchten, zeigen Sie dies der Experimentleitung bitte per Handzeichen an.

Instructions, General

Spielanleitung

Im Rahmen des Experiments werden Ihnen Entscheidungsprobleme vorgelegt, an denen jeweils zwei Personen beteiligt sind. Die Entscheidungsprobleme werden als Spiele und die daran beteiligten Personen als Spieler A und B bezeichnet.

Insgesamt werden Sie sechs Spiele gegen wechselnde Gegner spielen. In jedem Spiel wird Ihnen ein anderer Experimentteilnehmer als Gegner zugeordnet. Sie werden über die Problemstellung und die zu treffenden Entscheidungen informiert. Alle Experimentteilnehmer erhalten die gleichen Informationen.

Sie treffen Ihre Entscheidungen am Computer. Die Benutzung des Computers wird Ihnen erklärt. Ihre Entscheidungen werden per Computer an Ihren jeweiligen Gegenspieler weitergeleitet. Dieser erfährt nur Ihre Entscheidung, jedoch nicht Ihren Namen oder Ihre Teilnehmernummer, d.h. Ihre Entscheidungen sind anonym.

Alle Experimentteilnehmer sind in zwei Gruppen aufgeteilt; die Gruppe der Spieler A und die Gruppe der Spieler B.

Sie gehören zur Gruppe der Spieler A !

Das heißt, in allen sechs Spielen entscheiden Sie in der Rolle des Spielers A.

Spiel 1

Das Spiel gliedert sich in zwei Phasen, die als "Produktionsphase" und "Verteilungsphase" bezeichnet werden.

In der ersten Spielphase - der Produktionsphase - wählen beide Spieler eine Aktion - die "Produktionsmenge". Die möglichen Produktionsmengen sind in einer Tabelle angegeben (siehe Infoblatt "Erlösformel und Kostentabelle").

Jede Produktion verursacht Kosten. Die Kosten sind umso höher je größer die Produktionsmenge ist. Sie sind in einer Tabelle angegebenen (siehe Infoblatt "Erlösformel und Kostentabelle"). Jeder Spieler muß seine Kosten zahlen.

Aus den Aktionen beider Spieler ergibt sich ein gemeinsamer Erlös. Für diesen Erlös gilt die angegebene Erlösformel (siehe Infoblatt "Erlösformel und Kostentabelle"). Kosten und Erlöse sind in Pfennigen angegeben. Alle Spieler erhalten die gleiche Kostentabelle.

Instructions, Player A, 1

Spielanleitung

Im Rahmen des Experiments werden Ihnen Entscheidungsprobleme vorgelegt, an denen jeweils zwei Personen beteiligt sind. Die Entscheidungsprobleme werden als Spiele und die daran beteiligten Personen als Spieler A und B bezeichnet.

Insgesamt werden Sie sechs Spiele gegen wechselnde Gegner spielen. In jedem Spiel wird Ihnen ein anderer Experimentteilnehmer als Gegner zugeordnet. Sie werden über die Problemstellung und die zu treffenden Entscheidungen informiert. Alle Experimentteilnehmer erhalten die gleichen Informationen.

Sie treffen Ihre Entscheidungen am Computer. Die Benutzung des Computers wird Ihnen erklärt. Ihre Entscheidungen werden per Computer an Ihren jeweiligen Gegenspieler weitergeleitet. Dieser erfährt nur Ihre Entscheidung, jedoch nicht Ihren Namen oder Ihre Teilnehmernummer, d.h. Ihre Entscheidungen sind anonym.

Alle Experimentteilnehmer sind in zwei Gruppen aufgeteilt; die Gruppe der Spieler A und die Gruppe der Spieler B.

Sie gehören zur Gruppe der Spieler B !

Das heißt, in allen sechs Spielen entscheiden Sie in der Rolle des Spielers B.

Spiel 1

Das Spiel gliedert sich in zwei Phasen, die als Produktionsphase und Verteilungsphase bezeichnet werden.

In der ersten Spielphase - der Produktionsphase - wählen beide Spieler eine Aktion - die "Produktionsmenge". Die möglichen Produktionsmengen sind in einer Tabelle angegeben (siehe Infoblatt "Erlösformel und Kostentabelle").

Jede Produktion verursacht Kosten. Die Kosten sind umso höher je größer die Produktionsmenge ist. Sie sind in einer Tabelle angegebenen (siehe Infoblatt "Erlösformel und Kostentabelle"). Jeder Spieler muß seine Kosten zahlen.
Aus den Aktionen beider Spieler ergibt sich ein gemeinsamer Erlös. Für diesen Erlös gilt die angegebene Erlösformel (siehe Infoblatt "Erlösformel und Kostentabelle"). Kosten und Erlöse sind in Pfennigen angegeben. Alle Spieler erhalten die gleiche Kostentabelle.

Instructions, Player B, 1

In der zweiten Spielphase - der Verteilungsphase - kann der Erlös nach folgenden Regeln aufgeteilt werden:

Zunächst werden dem Spieler A die Produktionsmengen und der Erlös mitgeteilt. Spieler A macht einen Aufteilungsvorschlag; das heißt, er legt fest, wieviel er vom Erlös für sich fordert. Danach werden dem Spieler B die Produktionsmengen und der Erlös mitgeteilt sowie der Betrag, den Spieler A für sich fordert. Spieler B entscheidet dann, ob er die Forderung von Spieler A akzeptiert oder ablehnt. Falls er ablehnt, erhalten beide nichts, müssen aber dennoch ihre jeweiligen Kosten tragen. Falls er akzeptiert, bekommt Spieler A seinen Forderungsbetrag abzüglich seiner Kosten, während der B den restlichen Teil des Erlöses abzüglich seiner Kosten erhält.

Der A wird über die Entscheidung des Spielers B informiert. Somit sind am Ende beide über alle Entscheidungen informiert. Das Spiel ist beendet.

Kurzdarstellung des Spielablaufs

Nach diesen ausführlichen Hinweisen sind nachfolgend noch einmal die einzelnen Schritte des Spielablaufs und die Auszahlungsregeln aufgelistet. Diese Informationen werden während des Spieles am Bildschirm angezeigt.

1. Beide Spieler wählen ihre Produktionsmengen.
2. Spieler A wird per Computer über den Erlös (und die Produktionsmengen beider Spieler) informiert und macht einen Vorschlag zur Aufteilung des Erlöses; d.h. er bestimmt den Betrag, den er für sich fordert.
3. Spieler B wird per Computer über den Erlös (und die Produktionsmengen beider Spieler) sowie die Forderung des A informiert und entscheidet über Annahme oder Ablehnung der Forderung.
4. Spieler A wird über die Entscheidung des B informiert.

Auszahlungsregeln:

- Die Erlöse und Kosten ergeben sich gemäß dem Infoblatt "Erlösformel und Kostentabelle".
- Wenn B ablehnt, bekommen beide Spieler nichts ausgezahlt, müssen aber dennoch ihre jeweiligen Kosten tragen.
- Wenn B akzeptiert, bekommt Spieler A seine Forderung abzüglich seiner Kosten ausgezahlt. Spieler B erhält den restlichen Teil des Erlöses abzüglich seiner Kosten.

Spiel 2 bis Spiel 6

In den Spielen 2 bis 6 wird das gleiche Spiel gegen andere Gegner gespielt. Während des gesamten Experiments treffen Sie auf jeden Ihrer Gegenspieler nur je einmal.

Instructions, Players A and B, 2

Erlösformel und Kostentabelle

(Diese Informationen sind für alle Spieler gleich).

Erlösformel:

E = 22 Pfennig · Menge A + 44 Pfennig · Menge B

Kostentabelle:

Eigene Menge	Eigene Kosten (in Pfennig)
0	0
1	26
2	34
3	40
4	45
5	49
6	53
7	57
8	63
9	71
10	82
11	96
12	115
13	138
14	166
15	201
16	243
17	293
18	350
19	417
20	493
21	580
22	677
23	787
24	908
25	1043
26	1192
27	1355
28	1533
29	1727
30	1938

Instructions, Cost Table

General Information

For your participation within this experiment you receive 5,- DM as a gift. Depending on your success you can earn more money, but losses are possible as well. Your winnings will be paid, indeed. Losses will be collected from you or booked against your participation premia. In all games you have the option to avoid any financial risk.

For convenience all payments will be booked first onto a computer account and will be transferred later through the secrataries office of Professor Dr. Reinhard Tietz, room 110 C. You can pick up your money there throughout the semester on Wednesdays, 10 to 12 or 13 to 15 hours. Earnings which are not picked up by the end of winter semester 93/94 will not be paid out anymore.

Your participant-number is:						

You need this information sheet with your participant-number as identification card for receiving money. Take it with you at the end of the experiment and show it when you pick up your winnings.

We kindly ask you not to talk to one another throughout the experiment. If there is something you don't understand or if you have any other question, raise your arm to indicate this to the experimenter.

English Translation: Instructions, General

Introduction

In this experiment you will face decision problems in which two persons are involved. In the sequel the decision problems will be referred to as games, and the persons involved will be called players A and B.
Alltogether you will play six games against changing opponents. In each game a different person participating in this experiment will be assigned as your opponent at random. You will receive information upon the decision problem and the decisions you can make. All participants receive the same informations.
You enter your decisions into the computer. Handling of the computer will be explained to you. Your decisions will be transferred via the computer to your opponent. He will be informed about your decision but not about your name nor your participant number, i.e. your decisions are anonymous.

All participants of the experiment are divided into two groups, the group of players A and the group of players B.

You belong to the group of players A!

I.e., you will decide in the role of player A in all six games.

Game 1

The game is divided into two phases, referred to as production phase and distribution phase.

Within the first phase of the game - the production phase - both players choose an action - the "production quantity". The possible production quantities are given in a table (see information sheet "Return Formula and Cost Table").

Production causes cost. The cost are higher the larger the production quantity. They are denoted in a table (see information sheet "Return Formula and Cost Table"). Each player has to bear his cost.

The actions of both players determine a joint return according to the given return formula (see information sheet "Return Formula and Cost Table"). Cost and return are denoted in Pfennig. All players receive the same cost table.

English Translation: Instructions, Player A, 1

Introduction

In this experiment you will face decision problems in which two persons are involved. In the sequel the decision problems will be referred to as games, and the persons involved will be called players A and B.
Alltogether you will play six games against changing opponents. In each game a different person participating in this experiment will be assigned as your opponent at random. You will receive information upon the decision problem and the decisions you can make. All participants receive the same informations.
You enter your decisions into the computer. Handling of the computer will be explained to you. Your decisions will be transferred via the computer to your opponent. He will be informed about your decision but not about your name nor your participant number, i.e. your decisions are anonymous.

All participants of the experiment are divided into two groups, the group of players A and the group of players B.

You belong to the group of players B!

I.e., you will decide in the role of player B in all six games.

Game 1

The game is divided into two phases, referred to as production phase and distribution phase.

Within the first phase of the game - the production phase - both players choose an action - the "production quantity". The possible production quantities are given in a table (see information sheet "Return Formula and Cost Table").

Production causes cost. The cost are higher the larger the production quantity. They are denoted in a table (see information sheet "Return Formula and Cost Table"). Each player has to bear his cost.
The actions of both players determine a joint return according to the given return formula (see information sheet "Return Formula and Cost Table"). Cost and return are denoted in Pfennig. All players receive the same cost table.

English Translation: Instructions, Player B, 1

Within the second phase of the game - the distribution phase - the return can be divided according to the following rules:

First, player A will be informed about the chosen production quantities and the return. Player A suggests a distribution; i.e. he states how much of the return he demands for himself. Thereafter player B will be informed about the production quantities, the return and the amount demanded by player A. Then player B decides, whether he accepts or rejects the demand of player A. If he rejects, both get nothing, but nevertheless have to pay their respective cost. If he accepts, player A gets his demanded amount minus his cost, while B gets the rest of the return minus his cost.

A will be informed about the decision of player B. Thus, at the end of the game both are informed about all decisions. This is the end of the game.

Short Description of the Course of the Game

After these detailed informations in the following the various steps of the game and the payoff rules are listed. These informations are also available on screen.

1. Both players choose their production quantities.
2. Player A will be informed via the computer about the return (and the production quantities of both players) and suggests a distribution of the return; i.e. he states how much of the return he demands for himself.
3. Player B will be informed via the computer about the return (and the production quantities of both players) as well as the demand of player A and decides about acceptance or rejection of the demand.
4. Player A will be informed about the decision of B.

Payoff Rules:

- Return and cost are determined according to information sheet "Return Formula and Cost Table".
- If B rejects, both players get nothing, but nevertheless have to bear their respective cost.
- If B accepts, player A gets his demand minus his cost. Player B gets the rest of the return minus his cost.

Game 2 to Game 6

In games 2 to 6 the same game is played against different opponents. Within the complete experiment you meet each of your opponents only once.

English Translation: Instructions, Player A and B, 2

Return Formula and Cost Table

(These informations are the same for all players)

Return Formula:

R = 22 Pfennig · Quantity A + 44 Pfennig · Quantity B

Cost Table:

Own Quantity	Own Cost (in Pfennig)
0	0
1	26
2	34
3	40
4	45
5	49
6	53
7	57
8	63
9	71
10	82
11	96
12	115
13	138
14	166
15	201
16	243
17	293
18	350
19	417
20	493
21	580
22	677
23	787
24	908
25	1043
26	1192
27	1355
28	1533
29	1727
30	1938

English Translation: Instructions, Cost Table

Anweisung zum 16-PA

Auf den folgenden zwei Seiten finden Sie eine Liste von Persönlichkeitseigenschaften. Beschreiben Sie sich bitte selbst anhand dieser Liste, indem Sie *in jeder Zeile* auf der Skala ankreuzen (x), in welchem Ausmaß entweder die linke oder die rechte Eigenschaft auf Sie zutrifft. Auch wenn Sie mitunter sagen möchten, daß je nach Umständen beide Eigenschaften zutreffen oder daß Sie eigentlich keine Gegensätze darstellen, sollten Sie nur *einen* Wert ankreuzen und damit angeben, welche der beiden Eigenschaften eher zutrifft un in welchem Ausmaß sie zutrifft. Füllen sie bitte beide Seiten aus und übersehen Sie kein Eigenschaftspaar!

Da alle Angaben anonym sind, können Sie unbesorgt niederschreiben, was Sie wirklich von sich denken. Es ist niemand da, auf den Sie einen guten Eindruck machen müßten. Nur wenn Sie ganz ehrlich antworten, sind die Ergebnisse wissenschaftlich verwertbar. Wir bitten Sie daher, am Ende des Fragebogens anzugeben, wie sehr wir uns auf Ihre Angaben verlassen können.

Bitte geben Sie Alter, Geschlecht, höchsten Schulabschluß und heutiges Datum an:

Alter in Jahren	*Geschlecht*	*Schulabschluß*
☐ 0 – 19	☐ männlich	☐ Reifeprüfung
☐ 20 – 23	☐ weiblich	☐ Zwischenprüfung
☐ 24 – 27		☐ Diplomarbeit
☐ 28 – 31		☐ Examen
☐ 32 – 35		
☐ 34 – ...		

Heutiges Datum: ____ Tag ____ Monat ____ Jahr

Code: ☐☐☐☐☐☐

16-PA Personality Questionnaire, 1

16-PA

sachbezogen	4	3	2	1	0	1	2	3	4	kontaktfreudig
langsam im Denken	4	3	2	1	0	1	2	3	4	schnell im Denken
leicht zu beunruhigen	4	3	2	1	0	1	2	3	4	seelisch stabil
anpassungsbereit	4	3	2	1	0	1	2	3	4	eigenwillig
ruhig	4	3	2	1	0	1	2	3	4	lebhaft
sorglos	4	3	2	1	0	1	2	3	4	gewissenhaft
schüchtern	4	3	2	1	0	1	2	3	4	draufgängerisch
dickfellig	4	3	2	1	0	1	2	3	4	feinfühlig
gutgläubig	4	3	2	1	0	1	2	3	4	mißtrauisch
realistisch	4	3	2	1	0	1	2	3	4	träumerisch
gerade heraus	4	3	2	1	0	1	2	3	4	diplomatisch
mit mir zufrieden	4	3	2	1	0	1	2	3	4	an mir zweifelnd
am Gewohnten festhaltend	4	3	2	1	0	1	2	3	4	für Veränderung aufgeschlossen
anlehnungsbedürftig	4	3	2	1	0	1	2	3	4	eigenständig
unbeherrscht	4	3	2	1	0	1	2	3	4	diszipliniert
ausgeglichen	4	3	2	1	0	1	2	3	4	reizbar
angespannt	4	3	2	1	0	1	2	3	4	gelassen
experimentierfreudig	4	3	2	1	0	1	2	3	4	an Bewährtem orientiert
geübt im Nachdenken	4	3	2	1	0	1	2	3	4	ungeübt im Nachdenken
zartbesaitet	4	3	2	1	0	1	2	3	4	derb
ängstlich-besorgt	4	3	2	1	0	1	2	3	4	unbeschwert
phantasievoll	4	3	2	1	0	1	2	3	4	nüchtern
auf Selbstbehauptung bedacht	4	3	2	1	0	1	2	3	4	zur Unterordnung bereit
warmherzig	4	3	2	1	0	1	2	3	4	kühl
prinzipientreu	4	3	2	1	0	1	2	3	4	leichtlebig
kritisch	4	3	2	1	0	1	2	3	4	nachsichtig
gerne unabhängig	4	3	2	1	0	1	2	3	4	gerne mit anderen
kultiviert	4	3	2	1	0	1	2	3	4	natürlich
selbstbeherrscht	4	3	2	1	0	1	2	3	4	wechselhaft
selbstsicher	4	3	2	1	0	1	2	3	4	scheu
seelisch widerstandfähig	4	3	2	1	0	1	2	3	4	seelisch wenig belastbar
erlebnishungrig	4	3	2	1	0	1	2	3	4	stillebedürftig
Sie können sich auf meine Angaben verlassen	4	3	2	1	0	1	2	3	4	Sie können sich auf meine Angaben nicht verlassen

Bitte vergewissern Sie sich, daß Sie kein Eigenschaftspaar ausgelassen und in jeder Zeile nur eine Maßzahl angekreuzt haben!

16-PA Personality Questionnaire, 2

B. Experimental Instructions Chapter 3

This appendix contains the German instructions to the experiment described in chapter 3 and an English translation. Subjects received no separate instructions for repeated game 2 except for the cost table (see below). It was announced that there would be another sequence of six games (games 7 to 12) against a new opponent (compared to games 1 to 6, but not changing between games 7 to 12).

For the 16PA–postexperimental questionnaire see the appendix to chapter 1.

Allgemeine Hinweise

Für Ihre Teilnahme an diesem Experiment bekommen Sie 5,- DM geschenkt. In Abhängigkeit von Ihrem Erfolg können Sie zusätzliches Geld gewinnen, aber natürlich auch Geld verlieren. Ihre Gewinne werden Ihnen tatsächlich ausgezahlt. Verluste werden einkassiert bzw. mit Gewinnen aus anderen Spielen und Ihrer Teilnahmeprämie verrechnet. In allen Spielen haben Sie die Möglichkeit, so zu entscheiden, daß Sie jedes finanzielle Risiko vermeiden.

Zur Vereinfachung werden alle Zahlungen zunächst nur im Computer verbucht und später durch das Sekretariat von Herrn Professor Dr. Reinhard Tietz, Raum 110 C abgewickelt. Sie können sich Ihr Geld dort während des Semesters jeweils mittwochs, 10 bis 12 Uhr oder 13 bis 15 Uhr, abholen. Gewinne, welche bis Ende des WS 94/95 nicht abgeholt werden, werden nicht mehr ausgezahlt.

Ihre Teilnehmernummer lautet:

Sie benötigen dieses Hinweisblatt mit Ihrer Teilnehmernummer als Ausweis für die Geldauszahlung. Nehmen Sie es am Ende des Experiments mit nach Hause und bringen Sie es mit, wenn Sie Ihren Gewinn abholen.

Wir bitten Sie, während des Experimentes nicht miteinander zu reden. Wenn Sie etwas nicht verstehen oder sonstige Fragen stellen möchten, zeigen Sie dies der Experimentleitung bitte per Handzeichen an.

Instructions, General

Spielanleitung

Im Rahmen des Experiments werden Ihnen Entscheidungsprobleme vorgelegt, an denen jeweils zwei Personen beteiligt sind. Die Entscheidungsprobleme werden als Spiele und die daran beteiligten Personen als Spieler A und B bezeichnet.

Insgesamt werden Sie sechs Spiele gegen den selben Gegner spielen. Sie werden über die Problemstellung und die zu treffenden Entscheidungen informiert. Alle Experimentteilnehmer erhalten die gleichen Informationen.

Sie treffen Ihre Entscheidungen am Computer. Die Benutzung des Computers wird Ihnen erklärt. Ihre Entscheidungen werden per Computer an Ihren jeweiligen Gegenspieler weitergeleitet. Dieser erfährt nur Ihre Entscheidung, jedoch nicht Ihren Namen oder Ihre Teilnehmernummer, d.h. Ihre Entscheidungen sind anonym.

Alle Experimentteilnehmer sind in zwei Gruppen aufgeteilt; die Gruppe der Spieler A und die Gruppe der Spieler B.

Sie gehören zur Gruppe der Spieler A !

Das heißt, in allen sechs Spielen entscheiden Sie in der Rolle des Spielers A.

Spiel 1

Das Spiel gliedert sich in zwei Phasen, die als "Produktionsphase" und "Verteilungsphase" bezeichnet werden.

In der ersten Spielphase - der Produktionsphase - wählen beide Spieler eine Aktion - die "Produktionsmenge". Die möglichen Produktionsmengen sind in einer Tabelle angegeben (siehe Infoblatt "Erlösformel und Kostentabelle").

Jede Produktion verursacht Kosten. Die Kosten sind umso höher je größer die Produktionsmenge ist. Sie sind in einer Tabelle angegebenen (siehe Infoblatt "Erlösformel und Kostentabelle"). Jeder Spieler muß seine Kosten zahlen.

Aus den Aktionen beider Spieler ergibt sich ein gemeinsamer Erlös. Für diesen Erlös gilt die angegebene Erlösformel (siehe Infoblatt "Erlösformel und Kostentabelle"). Kosten und Erlöse sind in Pfennigen angegeben. Alle Spieler erhalten die gleiche Kostentabelle.

Instructions, Player A, 1

Spielanleitung

Im Rahmen des Experiments werden Ihnen Entscheidungsprobleme vorgelegt, an denen jeweils zwei Personen beteiligt sind. Die Entscheidungsprobleme werden als Spiele und die daran beteiligten Personen als Spieler A und B bezeichnet.

Insgesamt werden Sie sechs Spiele gegen den selben Gegner spielen. Sie werden über die Problemstellung und die zu treffenden Entscheidungen informiert. Alle Experimentteilnehmer erhalten die gleichen Informationen.

Sie treffen Ihre Entscheidungen am Computer. Die Benutzung des Computers wird Ihnen erklärt. Ihre Entscheidungen werden per Computer an Ihren jeweiligen Gegenspieler weitergeleitet. Dieser erfährt nur Ihre Entscheidung, jedoch nicht Ihren Namen oder Ihre Teilnehmernummer, d.h. Ihre Entscheidungen sind anonym.

Alle Experimentteilnehmer sind in zwei Gruppen aufgeteilt; die Gruppe der Spieler A und die Gruppe der Spieler B.

Sie gehören zur Gruppe der Spieler B !

Das heißt, in allen sechs Spielen entscheiden Sie in der Rolle des Spielers B.

Spiel 1

Das Spiel gliedert sich in zwei Phasen, die als "Produktionsphase" und "Verteilungsphase" bezeichnet werden.

In der ersten Spielphase - der Produktionsphase - wählen beide Spieler eine Aktion - die "Produktionsmenge". Die möglichen Produktionsmengen sind in einer Tabelle angegeben (siehe Infoblatt "Erlösformel und Kostentabelle").

Jede Produktion verursacht Kosten. Die Kosten sind umso höher je größer die Produktionsmenge ist. Sie sind in einer Tabelle angegebenen (siehe Infoblatt "Erlösformel und Kostentabelle"). Jeder Spieler muß seine Kosten zahlen.

Aus den Aktionen beider Spieler ergibt sich ein gemeinsamer Erlös. Für diesen Erlös gilt die angegebene Erlösformel (siehe Infoblatt "Erlösformel und Kostentabelle"). Kosten und Erlöse sind in Pfennigen angegeben. Alle Spieler erhalten die gleiche Kostentabelle.

Instructions, Player B, 1

In der zweiten Spielphase - der Verteilungsphase - kann der Erlös nach folgenden Regeln aufgeteilt werden:

Zunächst werden dem Spieler A die Produktionsmengen und der Erlös mitgeteilt. Spieler A macht einen Aufteilungsvorschlag; das heißt, er legt fest, wieviel er vom Erlös für sich fordert. Danach werden dem Spieler B die Produktionsmengen und der Erlös mitgeteilt sowie der Betrag, den Spieler A für sich fordert. Spieler B entscheidet dann, ob er die Forderung von Spieler A akzeptiert oder ablehnt. Falls er ablehnt, erhalten beide nichts, müssen aber dennoch ihre jeweiligen Kosten tragen. Falls er akzeptiert, bekommt Spieler A seinen Forderungsbetrag abzüglich seiner Kosten, während der B den restlichen Teil des Erlöses abzüglich seiner Kosten erhält.

Der A wird über die Entscheidung des Spielers B informiert. Somit sind am Ende beide über alle Entscheidungen informiert. Das Spiel ist beendet.

Kurzdarstellung des Spielablaufs

Nach diesen ausführlichen Hinweisen sind nachfolgend noch einmal die einzelnen Schritte des Spielablaufs und die Auszahlungsregeln aufgelistet. Diese Informationen werden während des Spieles am Bildschirm angezeigt.

1. Beide Spieler wählen ihre Produktionsmengen.
2. Spieler A wird per Computer über den Erlös (und die Produktionsmengen beider Spieler) informiert und macht einen Vorschlag zur Aufteilung des Erlöses; d.h. er bestimmt den Betrag, den er für sich fordert.
3. Spieler B wird per Computer über den Erlös (und die Produktionsmengen beider Spieler) sowie die Forderung des A informiert und entscheidet über Annahme oder Ablehnung der Forderung.
4. Spieler A wird über die Entscheidung des B informiert.

Auszahlungsregeln:

- Die Erlöse und Kosten ergeben sich gemäß dem Infoblatt "Erlösformel und Kostentabelle".
- Wenn B ablehnt, bekommen beide Spieler nichts ausgezahlt, müssen aber dennoch ihre jeweiligen Kosten tragen.
- Wenn B akzeptiert, bekommt Spieler A seine Forderung abzüglich seiner Kosten ausgezahlt. Spieler B erhält den restlichen Teil des Erlöses abzüglich seiner Kosten.

Spiel 2 bis Spiel 6

In den Spielen 2 bis 6 wird das gleiche Spiel gegen den selben Gegner gespielt.

Instructions, Player A and B, 2

Erlösformel und Kostentabelle

(Diese Informationen sind für alle Spieler gleich)

Erlösformel:

E = 22 Pfennig · Menge A + 44 Pfennig · Menge B

Kostentabelle:

Eigene Menge	Eigene Kosten (in Pfennig)
0	0
1	26
2	34
3	40
4	45
5	49
6	53
7	57
8	63
9	71
10	82
11	96
12	115
13	138
14	166
15	201
16	243
17	293
18	350
19	417
20	493
21	580
22	677
23	787
24	908
25	1043
26	1192
27	1355
28	1533
29	1727
30	1938

Cost Table, Repeated Game 1

Erlösformel und Kostentabelle

(Diese Informationen sind für alle Spieler gleich)

Erlösformel:

E = 22 Pfennig · Menge A + 44 Pfennig · Menge B

Kostentabelle:

Eigene Menge	Eigene Kosten (in Pfennig)	Grenzkosten* (in Pfennig)	Durchschnitts-kosten (in Pfennig)
0	0		
1	26	26	26,00
2	34	8	17,00
3	40	6	13,33
4	45	5	11,25
5	49	4	9,80
6	53	4	8,83
7	57	4	8,14
8	63	6	7,88
9	71	8	7,89
10	82	11	8,20
11	96	14	8,73
12	115	19	9,58
13	138	23	10,62
14	166	28	11,86
15	201	35	13,40
16	243	42	15,19
17	293	50	17,24
18	350	57	19,44
19	417	67	21,95
20	493	76	24,65
21	580	87	27,62
22	677	97	30,77
23	787	110	34,22
24	908	121	37,83
25	1043	135	41,72
26	1192	149	45,85
27	1355	163	50,19
28	1533	178	54,75
29	1727	194	59,55
30	1938	211	64,60

*Kostenänderung, die durch Erhöhung der Produktion um eine Mengeneinheit verursacht wurde.

Extended Cost Table, Repeated Game 1

Erlösformel und Kostentabelle für die Spiele 7 bis 12

(Diese Informationen sind für alle Spieler gleich)

Erlösformel:

E = 24 Pfennig · Menge A + 48 Pfennig · Menge B

Kostentabelle:

Eigene Menge	Eigene Kosten (in Pfennig)
0	0
1	79
2	85
3	89
4	93
5	96
6	99
7	102
8	107
9	113
10	121
11	131
12	145
13	163
14	184
15	210
16	242
17	279
18	322
19	372
20	429
21	494
22	567
23	649
24	740
25	842
26	953
27	1075
28	1209
29	1355
30	1513

Cost Table, Repeated Game 2

Erlösformel und Kostentabelle für die Spiele 7 bis 12

(Diese Informationen sind für alle Spieler gleich)

Erlösformel:

E = 24 Pfennig · Menge A + 48 Pfennig · Menge B

Kostentabelle:

Eigene Menge	Eigene Kosten (in Pfennig)	Grenzkosten* (in Pfennig)	Durchschnitts-kosten (in Pfennig)
0	0		
1	79	79	79,00
2	85	6	42,50
3	89	4	29,67
4	93	4	23,25
5	96	3	19,20
6	99	3	16,50
7	102	3	14,57
8	107	5	13,38
9	113	6	12,56
10	121	8	12,10
11	131	10	11,91
12	145	14	12,08
13	163	18	12,54
14	184	21	13,14
15	210	26	14,00
16	242	32	15,13
17	279	37	16,41
18	322	43	17,89
19	372	50	19,58
20	429	57	21,45
21	494	65	23,52
22	567	73	25,77
23	649	82	28,22
24	740	91	30,83
25	842	102	33,68
26	953	111	36,65
27	1075	122	39,81
28	1209	134	43,18
29	1355	146	46,72
30	1513	158	50,43

*Kostenänderung, die durch Erhöhung der Produktion um eine Mengeneinheit verursacht wurde.

Extended Cost Table, Repeated Game 2

General Information

For your participation within this experiment you receive 5,- DM as a gift. Depending on your success you can earn more money, but losses are possible as well. Your winnings will be paid, indeed. Losses will be collected from you or booked against your participation premia. In all games you have the option to avoid any financial risk.

For convenience all payments will be booked first onto a computer account and will be transferred later through the secrataries office of Professor Dr. Reinhard Tietz, room 110 C. You can pick up your money there throughout the semester on Wednesdays, 10 to 12 or 13 to 15 hours. Earnings which are not picked up by the end of winter semester 94/95 will not be paid out anymore.

Your participant-number is:

You need this information sheet with your participant-number as identification card for receiving money. Take it with you at the end of the experiment and show it when you pick up your winnings.

We kindly ask you not to talk to one another throughout the experiment. If there is something you don't understand or if you have any other question, raise your arm to indicate this to the experimenter.

English Translation: Instr., General

Introduction

In this experiment you will face decision problems in which two persons are involved. In the sequel the decision problems will be referred to as games, and the persons involved will be called players A and B.
Alltogether you will play six games against the same opponent. You will receive information upon the decision problem and the decisions you can make. All participants receive the same informations.
You enter your decisions into the computer. Handling of the computer will be explained to you. Your decisions will be transferred via the computer to your opponent. He will be informed about your decision but not about your name nor your participant number, i.e. your decisions are anonymous.

All participants of the experiment are divided into two groups, the group of players A and the group of players B.

You belong to the group of players A!

I.e., you will decide in the role of player A in all six games.

Game 1

The game is divided into two phases, referred to as production phase and distribution phase.

Within the first phase of the game - the production phase - both players choose an action - the "production quantity". The possible production quantities are given in a table (see information sheet "Return Formula and Cost Table").

Production causes cost. The cost are higher the larger the production quantity. They are denoted in a table (see information sheet "Return Formula and Cost Table"). Each player has to bear his cost.

The actions of both players determine a joint return according to the given return formula (see information sheet "Return Formula and Cost Table"). Cost and return are denoted in Pfennig. All players receive the same cost table.

English Translation: Instr., Player A, 1

Introduction

In this experiment you will face decision problems in which two persons are involved. In the sequel the decision problems will be referred to as games, and the persons involved will be called players A and B.
Alltogether you will play six games against the same opponent. You will receive information upon the decision problem and the decisions you can make. All participants receive the same informations.
You enter your decisions into the computer. Handling of the computer will be explained to you. Your decisions will be transferred via the computer to your opponent. He will be informed about your decision but not about your name nor your participant number, i.e. your decisions are anonymous.

All participants of the experiment are divided into two groups, the group of players A and the group of players B.

You belong to the group of players B!

I.e., you will decide in the role of player B in all six games.

Game 1

The game is divided into two phases, referred to as production phase and distribution phase.

Within the first phase of the game - the production phase - both players choose an action - the "production quantity". The possible production quantities are given in a table (see information sheet "Return Formula and Cost Table").

Production causes cost. The cost are higher the larger the production quantity. They are denoted in a table (see information sheet "Return Formula and Cost Table"). Each player has to bear his cost.
The actions of both players determine a joint return according to the given return formula (see information sheet "Return Formula and Cost Table"). Cost and return are denoted in Pfennig. All players receive the same cost table.

English Translation: Instr., Player B, 1

Within the second phase of the game - the distribution phase - the return can be divided according to the following rules:

First, player A will be informed about the chosen production quantities and the return. Player A suggests a distribution; i.e. he states how much of the return he demands for himself. Thereafter player B will be informed about the production quantities, the return and the amount demanded by player A. Then player B decides, whether he accepts or rejects the demand of player A. If he rejects, both get nothing, but nevertheless have to pay their respective cost. If he accepts, player A gets his demanded amount minus his cost, while B gets the rest of the return minus his cost.

A will be informed about the decision of player B. Thus, at the end of the game both are informed about all decisions. This is the end of the game.

Short Description of the Course of the Game

After these detailed informations in the following the various steps of the game and the payoff rules are listed. These informations are also available on screen.

1. Both players choose their production quantities.
2. Player A will be informed via the computer about the return (and the production quantities of both players) and suggests a distribution of the return; i.e. he states how much of the return he demands for himself.
3. Player B will be informed via the computer about the return (and the production quantities of both players) as well as the demand of player A and decides about acceptance or rejection of the demand.
4. Player A will be informed about the decision of B.

Payoff Rules:

- Return and cost are determined according to information sheet "Return Formula and Cost Table".
- If B rejects, both players get nothing, but nevertheless have to bear their respective cost.
- If B accepts, player A gets his demand minus his cost. Player B gets the rest of the return minus his cost.

Game 2 to Game 6

In games 2 to 6 the same game is played against the same opponent.

English Translation: Instr., Players A and B, 2

Return Formula and Cost Table

(These informations are the same for all players)

Return Formula:

R = 22 Pfennig · Quantity A + 44 Pfennig · Quantity B

Cost Table:

Own Quantity	Own Cost (in Pfennig)
0	0
1	26
2	34
3	40
4	45
5	49
6	53
7	57
8	63
9	71
10	82
11	96
12	115
13	138
14	166
15	201
16	243
17	293
18	350
19	417
20	493
21	580
22	677
23	787
24	908
25	1043
26	1192
27	1355
28	1533
29	1727
30	1938

English Translation: Cost Table for RG1 (for all other cost tables used in the experiment see above)

C. Experimental Instructions Chapter 4

This appendix contains the German version and an English translation of the instructions to the experiment described in chapter 4 as well as other material that was used within the session ("Nash–Ultimatum"); i.e., in the session in which the Nash demand game was played first and the ultimatum game second. The instructions, etc. for the session with the reversed order of play ("Ultimatum–Nash") are analogous to these.

For the 16PA–postexperimental questionnaire see the appendix to chapter 1.

Allgemeine Hinweise

Für Ihre Teilnahme an diesem Experiment bekommen Sie 20,- DM geschenkt. In Abhängigkeit von Ihrem Erfolg können Sie zusätzliches Geld gewinnen, aber auch Geld verlieren. Ihre Gewinne werden Ihnen tatsächlich ausgezahlt. Verluste werden mit Ihrer Teilnahmeprämie verrechnet.

Zur Vereinfachung werden alle Zahlungen später durch das Sekretariat von Herrn Professor Dr. Werner Güth, Raum 412 abgewickelt. Sie können sich Ihr Geld dort während des Semesters jeweils montags bis donnerstags, 10 bis 13 Uhr, abholen. Gewinne, welche bis Ende des SS 96 nicht abgeholt werden, werden nicht mehr ausgezahlt.

Ihre Teilnehmernummer lautet:

Sie benötigen dieses Hinweisblatt mit Ihrer Teilnehmernummer als Ausweis für die Geldauszahlung. Nehmen Sie es am Ende des Experiments mit nach Hause und bringen Sie es mit, wenn Sie Ihren Gewinn abholen.

Wir bitten Sie, während des Experimentes nicht miteinander zu reden. Verschiedene Teilnehmer können mit unterschiedlichen Situationen konfrontiert sein. Wenn Sie etwas nicht verstehen oder sonstige Fragen stellen möchten, zeigen Sie dies der Experimentleitung bitte per Handzeichen an.

Instructions, General Remarks

Spielanleitung

Im Rahmen des Experiments wird Ihnen ein Entscheidungsproblem vorgelegt, an dem jeweils zwei Personen beteiligt sind. Das Entscheidungsproblem wird nachfolgend als "Spiel" und die daran beteiligten Personen als "Spieler X und Y" bezeichnet. Innerhalb des Spiels werden alle Geldzahlungen in einer fiktiven Geldeinheit ECU (Experimental Currency Unit) notiert. Bei der Auszahlung der Experimentgelder erhalten Sie pro ECU jeweils 10 Pfennig.

Ihr Gegenspieler wird Ihnen zufällig zugeordnet. Er erfährt nur Ihre Entscheidung, jedoch nicht Ihren Namen oder Ihre Teilnehmernummer, d.h. Ihre Entscheidungen bleiben anonym.

Alle Experimentteilnehmer sind in zwei Gruppen aufgeteilt; die Gruppe der Spieler X und die Gruppe der Spieler Y.

Sie gehören zur Gruppe der Spieler X !

Das heißt, Sie entscheiden in der Rolle des Spielers X.

Das Spiel gliedert sich in zwei Phasen, die als "Produktionsphase" und "Verteilungsphase" bezeichnet werden.

In der sogenannten **Produktionsphase** wählen beide Spieler gleichzeitig ihre "individuelle Produktionsmenge", die mit x (für Spieler X) und y (für Spieler Y) bezeichnet werden. Es können nur ganzzahlige Produktionsmengen zwischen Null und 50 gewählt werden. Produktion verursacht Kosten in Höhe von 1 ECU je Mengeneinheit. Jeder Spieler muß seine individuellen Kosten - d.h. die Kosten seiner eigenen Produktionsmenge - tragen.

Aus den gewählten Mengen x und y ergibt sich ein gemeinsamer Erlös E, der in der zweiten Spielphase, der sogenannten **Verteilungsphase**, auf die Parteien X und Y aufgeteilt wird. Für den Erlös gilt folgende Berechnungsformel:

$$E = (3\text{ ECU} \cdot x) + (6\text{ ECU} \cdot y)$$

In der Verteilungsphase wird der Erlös nach folgenden Regeln aufgeteilt:

Zunächst wird beiden Spielern der zu verteilende Betrag E bekannt gegeben. Danach wählt jeder in Unkenntnis der Entscheidung des anderen einen Forderungsbetrag (F_X bzw. F_Y) und einen Betrag für seine Akzeptanzgrenze (A_X bzw. A_Y). Es muß gelten:

$$\text{Erlös} \geq \text{Forderungsbetrag} \geq \text{Akzeptanzgrenze} \geq \text{Null}$$

Hierbei können nur ganzzahlige ECU-Beträge gewählt werden. Die Experimentleitung sammelt die Entscheidungsformulare ein und errechnet die Auszahlungsbeträge gemäß den nachfolgenden Auszahlungsregeln.

Instructions, Game Rules, Player X

Spielanleitung

Im Rahmen des Experiments wird Ihnen ein Entscheidungsproblem vorgelegt, an dem jeweils zwei Personen beteiligt sind. Das Entscheidungsproblem wird nachfolgend als "Spiel" und die daran beteiligten Personen als "Spieler X und Y" bezeichnet. Innerhalb des Spiels werden alle Geldzahlungen in einer fiktiven Geldeinheit ECU (Experimental Currency Unit) notiert. Bei der Auszahlung der Experimentgelder erhalten Sie pro ECU jeweils 10 Pfennig.

Ihr Gegenspieler wird Ihnen zufällig zugeordnet. Er erfährt nur Ihre Entscheidung, jedoch nicht Ihren Namen oder Ihre Teilnehmernummer, d.h. Ihre Entscheidungen bleiben anonym.

Alle Experimentteilnehmer sind in zwei Gruppen aufgeteilt; die Gruppe der Spieler X und die Gruppe der Spieler Y.

Sie gehören zur Gruppe der Spieler Y !

Das heißt, Sie entscheiden in der Rolle des Spielers Y.

Das Spiel gliedert sich in zwei Phasen, die als "Produktionsphase" und "Verteilungsphase" bezeichnet werden.

In der sogenannten **Produktionsphase** wählen beide Spieler gleichzeitig ihre "individuelle Produktionsmenge", die mit x (für Spieler X) und y (für Spieler Y) bezeichnet werden. Es können nur ganzzahlige Produktionsmengen zwischen Null und 50 gewählt werden. Produktion verursacht Kosten in Höhe von 1 ECU je Mengeneinheit. Jeder Spieler muß seine individuellen Kosten - d.h. die Kosten seiner eigenen Produktionsmenge - tragen.

Aus den gewählten Mengen x und y ergibt sich ein gemeinsamer Erlös E, der in der zweiten Spielphase, der sogenannten **Verteilungsphase**, auf die Parteien X und Y aufgeteilt wird. Für den Erlös gilt folgende Berechnungsformel:

$$E = (3 \text{ ECU} \cdot x) + (6 \text{ ECU} \cdot y)$$

In der Verteilungsphase wird der Erlös nach folgenden Regeln aufgeteilt:

Zunächst wird beiden Spielern der zu verteilende Betrag E bekannt gegeben. Danach wählt jeder in Unkenntnis der Entscheidung des anderen einen Forderungsbetrag (F_X bzw. F_Y) und einen Betrag für seine Akzeptanzgrenze (A_X bzw. A_Y). Es muß gelten:

$$\text{Erlös} \geq \text{Forderungsbetrag} \geq \text{Akzeptanzgrenze} \geq \text{Null}$$

Hierbei können nur ganzzahlige ECU-Beträge gewählt werden. Die Experimentleitung sammelt die Entscheidungsformulare ein und errechnet die Auszahlungsbeträge gemäß den nachfolgenden Auszahlungsregeln.

Instructions, Game Rules, Player Y

Auszahlungsregeln

Es gibt zwei Arten, auf die eine Einigung zustande kommen kann, die sogenannte F-Einigung und die sogenannte A-Einigung. Die resultierenden Auszahlungen sowie diejenigen im Falle der Nicht-Einigung werden nun im einzelnen aufgeführt.

Eine **F-Einigung** kommt zustande, falls

$$F_x + F_y \leq E$$

(d.h. falls die Summe der Forderungsbeträge beider Spieler den Erlösbetrag nicht übersteigt).

- X bekommt $F_x - x$
  (d.h. seinen Forderungsbetrag abzüglich seiner Kosten)
- Y bekommt $F_y - y$
  (d.h. seinen Forderungsbetrag abzüglich seiner Kosten)

Ein positiver Restbetrag $E - F_x - F_y$ verfällt.

Falls es zu keiner F-Einigung kam, dann kommt eine **A-Einigung** zustande, falls

$$A_x + A_y \leq E$$

(d.h. falls die Summe der Akzeptanzgrenzen beider Spieler den Erlösbetrag nicht übersteigt).

- X bekommt $A_x - x$
  (d.h. seine Akzeptanzgrenze abzüglich seiner Kosten)
- Y bekommt $A_y - y$
  (d.h. seine Akzeptanzgrenze abzüglich seiner Kosten)

Ein positiver Restbetrag $E - A_x - A_y$ verfällt.

Kommt es weder zur F-Einigung noch zur A-Einigung, tritt der **Nicht-Einigungsfall** ein. Der Erlös ist damit verloren. Dennoch muß jeder Spieler seine Kosten tragen. Das heißt,

- X trägt einen Verlust in Höhe von $-x$
- Y trägt einen Verlust in Höhe von $-y$

Instructions, Payoff Rules Game 1, Both Players

Ihre Teilnehmernummer: | | | | | | |

Kontrollfragen

Wir bitten Sie nun, uns folgende Fragen zu beantworten, anhand derer wir prüfen, ob die Instruktionen vollständig verstanden wurden. Bei unvollständiger, nicht aber bei unrichtiger Beantwortung müssen wir Sie leider vom Experiment ausschließen.

Gehen Sie davon aus, daß X genau $\boldsymbol{x = 40}$ und Y genau $\boldsymbol{y = 30}$ gewählt hat.

a) Wie hoch ist der Gesamtbetrag, den X und Y untereinander aufteilen können?

Ferner sollen X und Y in den Verhandlungen $F_X = 150$, $A_X = 110$ bzw. $F_Y = 200$, $A_Y = 120$ wählen.

b) Kommt es zu

einer F-Einigung
oder zu einer A-Einigung
(bitte Zutreffendes ankreuzen)

c) Welche Beträge erhalten X und Y abzüglich ihrer Kosten?

X erhält

Y erhält

Control Questionnarire, Nash Game (analogous for ultimatum game)

Spiel 2

Es wird Ihnen ein neuer Gegenspieler zugelost. Es gilt eine neue Auszahlungsregel (siehe unten "Auszahlungsregel für Spiel 2"). Ansonsten gelten dieselben Regeln wie im ersten Spiel.

Auszahlungsregel für Spiel 2

Es wird per Zufallszug entschieden, welcher der beiden Spieler der sogenannte "Forderungsspieler" bzw. der sogenannte "Akzeptanzspieler" ist. Beide Fälle sind gleich wahrscheinlich. Die Auszahlungsregeln sind für beide Fälle analog. Dennoch werden der Vollständigkeit halber beide beschrieben.

Fall 1: "X ist Forderungsspieler, Y ist Akzeptanzspieler"

Die **Forderung F_x ist akzeptiert**, falls

$$E - F_x \geq A_y$$

(d.h. wenn X dem Y mindestens dessen Akzeptanzgrenze zugesteht).

- X bekommt $F_x - x$
  (d.h. seinen Forderungsbetrag abzüglich seiner Kosten)
- Y bekommt $E - F_x - y$
  (d.h. den Restbetrag abzüglich seiner Kosten)

Die **Forderung F_x ist abgelehnt**, falls

$$E - F_x < A_y$$

(d.h. wenn X dem Y weniger als dessen Akzeptanzgrenze zugesteht). Der Erlös ist damit verloren. Dennoch muß jeder Spieler seine Kosten tragen. Das heißt,

- X erleidet einen Verlust in Höhe von $-x$
- Y erleidet einen Verlust in Höhe von $-y$

Fall 2: "Y ist Forderungsspieler, X ist Akzeptanzspieler"

Die **Forderung F_y ist akzeptiert**, falls

$$E - F_y \geq A_x$$

(d.h. wenn Y dem X mindestens dessen Akzeptanzgrenze zugesteht).

- Y bekommt $F_y - y$
  (d.h. seinen Forderungsbetrag abzüglich seiner Kosten)
- X bekommt $E - F_y - x$
  (d.h. den Restbetrag abzüglich seiner Kosten)

Die **Forderung F_y ist abgelehnt**, falls

$$E - F_y < A_x$$

(d.h. wenn Y dem X weniger als dessen Akzeptanzgrenze zugesteht). Der Erlös ist damit verloren. Dennoch muß jeder Spieler seine Kosten tragen. Das heißt,

- Y erleidet einen Verlust in Höhe von $-y$
- X erleidet einen Verlust in Höhe von $-x$

Instructions, Payoff Rules Game 2, Both Players

Ihre Teilnehmernummer: | | | | | | |

Bitte überprüfen Sie, ob diese Teilnehmernummer mit jener auf dem Blatt „Allgemeine Hinweise" übereinstimmt.

Feld 1: Ihre Entscheidung (bitte eintragen):

Ich wähle eine Produktionsmenge von .... Einheiten.
(Es können nur ganzzahlige Mengen zwischen 0 und 50 gewählt werden.)

Feld 2: wird von der Experimentleitung ausgefüllt

Der gemeinsame Erlös beträgt .... ECU.

Feld 3: Ihre Entscheidung (bitte eintragen):

Meine Forderung beträgt .... ECU.

Meine Akzeptanzgrenze beträgt .... ECU.

Feld 4: wird von der Experimentleitung ausgefüllt

Es kam zu einer F-Einigung.
Es kam zu einer A-Einigung.

Sie erhalten einen Gewinn von .... ECU.

Sie haben einen Verlust von .... ECU.

Transfer Sheet, Nash Game (analogous in ultimatum game)

General Information

For your participation within this experiment you receive 20,- DM as a gift. Depending on your success you can earn more money, but losses are possible as well. Your wins will be paid, in fact. Losses will be booked against your participation premia.

For convenience all payments will be made through the secrataries office of Professor Dr. Werner Güth, room 412. You can pick up your money there throughout the semester on Tuesdays, Wednesdays or Thursdays between 10 and 13 hour. Earnings which are not picked up by the end of summer semester 1996 will not be paid out anymore.

Your participant-number is: | | | | | | |

You need this information sheet with your participant-number as identification card for receiving money. Take it with you at the end of the experiment and show it when you pick up your winnings.

We kindly ask you not to talk to one another throughout the experiment. Different particpants might be confronted with different situations. If there is something you do not understand or if you have any question, raise your arm to indicate it to the experimenter.

Translation: Instructions, General Remarks

Introduction

In this experiment you will face a decision problem where two persons are involved. The decision problem will subsequently be denoted as "game", and the persons participating in it will be denoted as "players X and Y". Within the game, all monetary payments are noted in a fictitious monetary unit ECU (Experimental Currency Unit). When you will be paid the earnings, you will receive 10 Pfennig per ECU.

You will be randomly matched with an opponent. He will only be informed about your decision, but not about your name or your participation number, i.e., your decision will be anonymous.

All participants of the experiment are divided into two groups, the group of players X and the group of players Y.

You belong to the group of players X!

I.e., you will decide in the role of part of player X.

The game is divided into two phases, denoted as "production phase" and "bargaining phase".

Within the so-called **production phase** the two players simultaneously choose the "individual production quantity", denoted by x (for player X) and y (for player Y). Only integers between 0 and 50 can be chosen as production quantities. Production causes costs of 1 ECU per quantity unit. Each player has to bear his individual cost - i.e. the cost of his own production quantity.

The chosen quantities x and y determine a joint return E, which will be distributed between players X and Y in the second phase of the game, the so-called **bargaining phase**. The return is calculated by the following formula:

$$E = (3 \text{ ECU} \cdot x) + (6 \text{ ECU} \cdot y)$$

In the bargaining phase, the return will be distributed according to the following rules:

Firstly, both players are informed about the amount E to be distributed. Then, each of them chooses an amount of demand (F_X and F_Y, resp.) and an amount for his acceptance bound (A_X and A_Y, resp.) without knowing the decision of the opponent. The following inequality must hold:

return ≥ demand ≥ acceptance bound ≥ zero

Only integer amounts of ECU can be chosen. The experimentators will collect the decision forms and compute the payoffs according to the following payoff rules.

Translation: Instructions, Game Rules, Player X

Introduction

In this experiment you will face a decision problem where two persons are involved. The decision problem will subsequently be denoted as "game", and the persons participating in it will be denoted as "players X and Y". Within the game, all monetary payments are noted in a fictitious monetary unit ECU (Experimental Currency Unit). When you will be paid the earnings, you will receive 10 Pfennig per ECU.

You will be randomly matched with an opponent. He will only be informed about your decision, but not about your name or your participation number, i.e., your decision will be anonymous.

All participants of the experiment are divided into two groups, the group of players X and the group of players Y.

You belong to the group of players Y!

I.e., you will decide in the role of part of player Y.

The game is divided into two phases, denoted as "production phase" and "bargaining phase".

Within the so-called **production phase** the two players simultaneously choose the "individual production quantity", denoted by x (for player X) and y (for player Y). Only integers between 0 and 50 can be chosen as production quantities. Production causes costs of 1 ECU per quantity unit. Each player has to bear his individual cost - i.e. the cost of his own production quantity.

The chosen quantities x and y determine a joint return E, which will be distributed between players X and Y in the second phase of the game, the so-called **bargaining phase**. The return is calculated by the following formula:

$$E = (3 \text{ ECU} \cdot x) + (6 \text{ ECU} \cdot y)$$

In the bargaining phase, the return will be distributed according to the following rules:

Firstly, both players are informed about the amount E to be distributed. Then, each of them chooses an amount of demand (F_X and F_y, resp.) and an amount for his acceptance bound (A_X and A_y, resp.) without knowing the decision of the opponent. The following inequality must hold:

return $\geq$ demand $\geq$ acceptance bound $\geq$ zero

Only integer amounts of ECU can be chosen. The experimentators will collect the decision forms and compute the payoffs according to the following payoff rules.

Translation: Instructions, Game Rules, Player Y

Payoff rules

There are two ways in which an agreement can be reached: the so-called F-agreement and the A-agreement. The payoffs resulting from an agreement as well as those in case of a disagreement will be listed below.

An **F-agreement** occurs, if

$$F_X + F_Y \leq E$$

(i.e. if the sum of the demands of both players does not exceed the return).

- X receives $F_X - x$
  (i.e. his demand minus his costs)
- Y receives $F_Y - y$
  (i.e. his demand minus his costs)

A positive rest E - F_X - F_Y is lost.

If no F-agreement has been reached, then an **A-agreement** is occurs, if

$$A_X + A_Y \leq E$$

(i.e. if the sum of the acceptanc bounds of both players does not exceed the return).

- X receives $A_X - x$
  (i.e. his acceptance bound minus his costs)
- Y receives $A_Y - y$
  (i.e. his acceptance bound minus his costs)

A positive rest E - A_X - A_Y is lost.

If there is neither an F-agreement nor an A-agreement, a **disagreement** occurs. In this case, the return is lost. However, each player must bear his costs. I.e.,

- X bears a loss of $-x$
- Y bears a loss of $-y$.

Translation: Instr., Payoff Rules, Game 1, Both Players

Your participation number: | | | | | | |

Control questions

You are now asked to answer the following questions which will help us to see if you have completely understood the instructions. In case of incomplete answers - not however in case of incorrect answers - you will be excluded from the participation in the experiment.

Suppose that X chose x = 40 and Y chose y = 30.

a) What is the amount of the joint return, on which X and Y can bargain?

Moreover, suppose that X and Y chose $F_X = 150$ and $A_X = 110$, $F_y = 200$ and $A_y = 120$, respectively.

b) Will there be a
F-agreement
or an A-agreement
(please tick appropriately)

c) Which payoff will X and Y receive net of their costs?

X receives

Y receives

Translation: Control Questionnaire, Nash Game (analogous for ultimatum game)

Game 2

A new opponent will be assigned to you at random. From now on, there are new payoff rules (see "Payoff rules for game 2" below). The other rules remain the same as in the first game.

Payoff rules for game 2

It will be randomly decided, which of the two players will be the so-called "demand player" and which the "acceptance player". Both cases are equally likely. Although the payoff rules are analogous for the two cases, both of them will be described for the sake of completeness.

Case 1: "X is the demand player, Y is the acceptance player"

The demand F_X is accepted, if

$$E - F_X \geq A_y$$

(i.e. if X concedes Y at least Y's acceptance bound).

- X receives $F_X - x$
  (i.e. his demand minus his costs)
- Y receives $E - F_X - y$
  (i.e. the rest of the amount minus his costs).

The demand F_X is rejected, if

$$E - F_X < A_y$$

i.e. if X concedes Y less than Y's acceptance bound. The return is then lost. However, each player will have to bear his costs, i.e.

- X bears a loss of $-x$
- Y bears a loss of $-y$.

Case 2: "Y is the demand player, X is the acceptance player"

The demand F_y is accepted, if

$$E - F_y \geq A_X$$

(i.e. if Y concedes X at least X's acceptance bound).

- Y receives $F_y - y$
  (i.e. his demand minus his costs)
- X receives $E - F_y - x$
  (i.e. the rest of the amount minus his costs).

The demand F_y is rejected, if

$$E - F_y < A_X$$

i.e. if Y concedes X less than X's acceptance bound. The return is then lost. However, each player will have to bear his costs, i.e.

- X has a loss of $-x$
- Y has a loss of $-y$.

Translation: Instr., Payoff Rules Game 2, Both Players

Your participation number: | | | | | | |

Please check if this participation number corresponds to the one on the form of the "General Instructions".

Field 1: Your decision (please fill in):

I choose a production quantity of .... units.
(You can only choose integer numbers between 0 and 50.)

Field 2: will be filled in by the experimenters

The joint return amounts to .... ECU.

Field 3: Your decision (please fill in):

My demand is .... ECU.

My acceptance bound is .... ECU.

Field 4: will be filled in by the experimenters

There was an F-agreement.
There was an A-agreement.

You receive a win of .... ECU.

Your loss is .... ECU.

Translation: Transfer Sheet, Nash Game (analogous in ultimatum game)

D. Mathematical Supplement To Chapter 5

D.1 Properties of the functions ΔS_1, ΔU_1^1 and ΔU_1^2

The type of equilibrium that is played in a given match (α, β) depends on the signs of ΔS_1, ΔU_1^1 and ΔU_1^2, which are functions of the four parameters of the model. The following paragraphs present some properties of these functions, which are important for the analysis. Within the evolutionary analysis it will be distinguished between match (α, β) and match (β, α). Of course, the properties of the functions for both matches are analogous to each other — only α and β switch — such that those for match (β, α) won't be given.

$$\Delta S_1(\alpha, \beta) = \underline{s}_1 - \bar{s}_1 = \frac{\alpha(1+c_1)-1}{2\alpha-1} - \frac{\beta(1-c_2)}{2\beta-1} \tag{D.1}$$

For all $(\alpha, \beta) \in \left(\frac{1}{2}, 1\right]^2 \wedge (c_1, c_2) \in [0,1)^2$: $\Delta S_1(\alpha, \beta)$ is strictly increasing in α, β, c_1 and c_2.

$$\Delta S_1(\alpha, \alpha) = \frac{\alpha(c_1+c_2)-1}{2\alpha-1} \tag{D.2}$$

Thus $c_1 + c_2 \leq 1$ implies that $\Delta S_1(\alpha, \alpha) \leq 0$ for all $\alpha \in \left[\frac{1}{2}, 1\right]$.

$$\begin{aligned} \Delta U_1^1(\alpha, \beta) &= U_1\left(\sigma_{10}^*(\alpha, \beta)\right) - U_1\left(\sigma_{01}^*(\alpha, \beta)\right) \\ &= \alpha(1-c_1) - \left(\bar{s}_1(2\alpha-1) + (1-\alpha)(1-c_2)\right) \\ &= \alpha(1-c_1) - (1-c_2)\frac{\alpha+\beta-1}{2\beta-1} \end{aligned} \tag{D.3}$$

For all $(\alpha, \beta) \in \left(\frac{1}{2}, 1\right]^2 \wedge (c_1, c_2) \in [0,1)^2$: $\Delta U_1^1(\alpha, \beta)$ is strictly decreasing in c_1 and strictly increasing in β and c_2. It is increasing in α, if $c_1 + c_2 - 2\beta c_1 + 2\beta - 2 \geq 0 \Leftrightarrow \beta \geq \frac{2-c_1-c_2}{2(1-c_1)}$.

$$\Delta U_1^1(\alpha,\alpha) = \alpha(1-c_1) - (1-c_2) \tag{D.4}$$

For all $\alpha \in \left[\frac{1}{2},1\right] \wedge (c_1,c_2) \in [0,1)^2$: $\Delta U_1^1(\alpha,\alpha)$ is strictly increasing in α, $\Delta U_1^1(\alpha,\alpha) = 0 \Leftrightarrow \alpha = \frac{1-c_2}{1-c_1}$ and $c_2 \le c_1 \Rightarrow \Delta U_1^1(\alpha,\alpha) \le 0$.

$$\begin{aligned}\Delta U_1^2(\alpha,\beta) &= U_1(\sigma_{10}^*(\alpha,\beta)) - U_1(\sigma_{11}^*(\alpha,\beta)) \\ &= \alpha(1-c_1) - (\alpha(\bar{s}_1 \cdot 2 - c_1) + (1-\alpha)((1-\bar{s}_1)\cdot 2 - c_2)) \\ &= \frac{2\beta(\alpha+\alpha c_2 - 1) + \alpha(c_2-3) - c_2 + 2}{2\beta - 1}\end{aligned} \tag{D.5}$$

For all $(\alpha,\beta) \in \left(\frac{1}{2},1\right]^2 \wedge c_2 \in [0,1)$: $\Delta U_1^2(\alpha,\beta)$ is strictly increasing in β and c_2. It is increasing in α, if $2\beta c_2 + 2\beta + c_2 - 3 \ge 0 \Longleftrightarrow \beta \ge \frac{3-c_2}{2(1+c_2)}$

$$\Delta U_1^2(\alpha,\alpha) = \frac{2\alpha(\alpha+\alpha c_2 - 1) + \alpha(c_2-3) - c_2 + 2}{2\alpha - 1} \tag{D.6}$$

For all $\alpha \in \left(\frac{1}{2},1\right] \wedge c_2 \in [0,1)$: $\Delta U_1^2(\alpha,\alpha)$ is strictly increasing in α and $\Delta U_1^2(\alpha,\alpha) = 0$ if and only if $\alpha = \frac{2-c_2}{1+c_2}$. Furthermore, $c_2 \le \frac{1}{2}$ implies $\Delta U_1^2(\alpha,\alpha) \le 0 \ \forall \alpha \in \left[\frac{1}{2},1\right]$.

D.2 Mutations $\beta < \alpha$ are Non–Improving if Candidate α Supports Efficiency

(This is the proof of Lemma 5.3)

Proof. Since α supports $(\sigma^*_{11}, \sigma^*_{11})$: $\Delta S_1(\alpha, \alpha) \leq 0$ and $\Delta U^2_1(\alpha, \alpha) \leq 0$ hold. Consider match (α, β). As $\Delta S_1(\alpha, \beta)$ and $\Delta U^2_1(\alpha, \beta)$ are increasing in β, the following holds for all $\beta < \alpha$:

$$\Delta S_1(\alpha, \beta) \leq \Delta S_1(\alpha, \alpha) \leq 0,$$
$$\Delta U^2_1(\alpha, \beta) \leq \Delta U^2_1(\alpha, \alpha) \leq 0.$$

Thus all mutations $\beta < \alpha$ induce play $\sigma^*_{11}(\alpha, \beta)$.

Now consider match (β, α). Since $\Delta S_1(\beta, \alpha)$ is increasing in β: $\Delta S_1(\alpha, \beta) \leq \Delta S_1(\alpha, \alpha) \leq 0$ for all $\beta < \alpha$. Thus, play in match (β, α) depends on $\Delta U^2_1(\beta, \alpha)$, which is monotonic in β, but either increasing or decreasing depending on the parameters. Suppose it is increasing, then $\Delta U^2_1(\beta, \alpha) \leq \Delta U^2_1(\alpha, \alpha) \leq 0$ for all $\beta < \alpha$. Alternatively, suppose it is decreasing. Then $\Delta U^2_1(\beta, \alpha)$ is maximal for $\beta = \frac{1}{2}$. In this case:

$$\Delta U^2_1(\beta, \alpha) = \frac{1}{2\alpha - 1}(1 - c_2)(\frac{1}{2} - \alpha) \leq 0.$$

Thus, all mutations $\beta < \alpha$ induce play $\sigma^*_{11}(\beta, \alpha)$.

So, every mutation $\beta < \alpha$ results in $(\sigma^*_{11}, \sigma^*_{11})$ and by lemma 5.1:

$$\Pi(\beta \mid \alpha \times \beta) < \Pi(\alpha \mid \alpha \times \alpha).$$

D.3 Technology 2.a

The following lemmata characterize evolutionary results for different subsets of the technology space. In many cases the statements in the proofs are conditional on a certain parameter range. In cases were the range is obvious — e.g. when it is contained in the definition of the technology subset or if it holds for a whole paragraph — it won't be repeated in the single statements. Also, in some cases the arguments for different subsets are equivalent such that a cross reference substitutes for repeating the statements. There are more cases where this could have been possible, but I refrained from doing so to improve clarity.

Lemma D.1 *For all technologies* $(c_1, c_2) \in T_{2.a}$*:*

$$\begin{array}{c} \alpha^2 = \frac{1}{c_1+c_2} \text{ sustains } (\sigma^*_{11}, \sigma^*_{11}) \text{ and is an ESS if} \\ 1 - c_2 - (2 - c_1 - c_2)^2 < 0, \end{array} \tag{D.7}$$

$$\begin{array}{c} \alpha^0 \text{ sustains } (\sigma^*_{01}, \sigma^*_{01}) \text{ and is an ESS if} \\ 1 - \frac{1}{2}c_2 < c_1. \end{array} \tag{D.8}$$

Proof. Conditions 5.12, 5.13 and lemma 5.2 imply that α^2 $(\Rightarrow (\sigma^*_{11}, \sigma^*_{11}))$ and α^0 $(\Rightarrow (\sigma^*_{01}, \sigma^*_{01}))$ are ESS-candidates.

Candidate: α^2

By lemma 5.3 it suffices to consider mutations $\beta > \alpha^2$.

Feasible Plays, Match (α, β): Since $\Delta S_1(\alpha^2, \alpha^2) = 0$ and $\Delta S_1(\alpha, \beta)$ strictly increasing in β:

$$\Delta S_1(\alpha^2, \beta) > 0 \forall \beta \in \left(\alpha^2, 1\right]. \tag{D.9}$$

Since

$$\Delta U_1^1(\alpha^2, \beta = 1) = \frac{c_2 - c_1}{c_1 + c_2} \leq 0 \forall (c_1, c_2) \in T_{2.a} \tag{D.10}$$

and $\Delta U_1^1(\alpha, \beta)$ increasing in β, it follows:

$$\Delta U_1^1(\alpha^2, \beta) \leq 0 \forall \beta \in \left[\alpha^2, 1\right]. \tag{D.11}$$

By D.9 and D.10 all mutations $\beta > \alpha^2$ induce play $\sigma^*_{01}(\alpha, \beta)$.

Feasible Plays, Match (β, α): Since $\Delta S_1(\alpha^2, \alpha^2) = 0$ and $\Delta S_1(\beta, \alpha)$ strictly increasing in β:

$$\Delta S_1(\beta,\alpha^2) > 0 \forall \beta \in \left(\alpha^2,1\right]. \tag{D.12}$$

Furthermore, $\Delta U_1^1(\alpha^2,\alpha^2) \leq 0$ (D.11). $\Delta U_1^1(\beta,\alpha^2)$ is decreasing in β if and only if

$$\alpha^2 \leq \frac{2-c_1-c_2}{2(1-c_1)}, \tag{D.13}$$

which is satisfied since $c_2 \leq c_1$. By D.12 and D.13 all mutations $\beta > \alpha^2$ induce play $\sigma_{01}^*(\beta,\alpha^2)$.

So, all mutations $\beta > \alpha^2$ imply SPE production vector $(\sigma_{01}^*, \sigma_{01}^*)$such that $\beta = 1$ is the challenger mutation (lemma 5.3). It remains to check whether $\beta = 1$ can be improving. If

$$1-c_2-(2-c_1-c_2)^2 < 0 \Longleftrightarrow \Pi(\beta \mid \alpha^2 \times \beta) < \Pi(\alpha^2 \mid \alpha^2 \times \alpha^2),$$

α^2 is an ESS. This proves D.7.

Candidate: α^0

Feasible Plays, Match (α,β): Note that

$$\Delta S_1(\alpha^0,\beta) \leq 0 \Longleftrightarrow \beta \leq \frac{c_1}{2c_1+c_2-1}, \tag{D.14}$$

$$\Delta U_1^2(\alpha^0,\beta) \leq 0 \forall \beta \leq 1, \text{ since } c_2 \leq \tfrac{1}{2}, \tag{D.15}$$

and

$$\Delta U_1^1(\alpha^0,\beta) \leq 0 \forall \beta \leq 1, \tag{D.16}$$

since $\Delta U_1^1(\alpha^0,\alpha^0) \leq 0$ and $\Delta U_1^1(\alpha,\beta)$ increasing in β. Thus play in match (α,β) is σ_{11}^* for mutations $\beta \leq \frac{c_1}{2c_1+c_2-1}$ and σ_{01}^* otherwise.

Feasible Plays, Match (β,α): Firstly,

$$\Delta S_1(\beta,\alpha^0) \leq 0 \Longleftrightarrow \beta \leq \frac{c_2}{2c_2+c_1-1} \text{ and} \tag{D.17}$$

$$\Delta U_1^2(\beta,\alpha^0) \leq 0 \Longleftrightarrow \beta(3c_2-1)-c_2 \leq 0, \tag{D.18}$$

which is satisfied $\forall \beta \in \left[\frac{1}{2},1\right]$ given $(c_1,c_2) \in T_{2.a}$. Furthermore

$$\Delta U_1^1(\beta,\alpha^0) \leq 0 \forall \beta \in \left[\frac{1}{2},1\right] \tag{D.19}$$

since $c_2 \leq c_1$. Thus play in match (β,α) is σ_{11}^* for mutations $\beta \leq \frac{c_2}{2c_2+c_1-1}$ and σ_{01}^* otherwise.

For all $(c_1,c_2) \in T_{2.a}$:

$$\frac{c_2}{2c_2+c_1-1} \leq \frac{c_1}{2c_1+c_2-1}$$

such that the following SPE production vectors are feasible via mutation:

$$\begin{array}{ll} (\sigma^*_{11}, \sigma^*_{11}) & \text{if } \beta \leq \frac{c_2}{2c_2+c_1-1} \\ (\sigma^*_{11}, \sigma^*_{01}) & \text{if } \frac{c_2}{2c_2+c_1-1} < \beta \leq \frac{c_1}{2c_1+c_2-1} \\ (\sigma^*_{01}, \sigma^*_{01}) & \text{otherwise} \end{array} \tag{D.20}$$

Mutations inducing $(\sigma^*_{01}, \sigma^*_{01})$ cannot be improving (lemma 5.1), here. Thus $\beta = \frac{c_2}{2c_2+c_1-1}$ as well as $\beta = \frac{c_1}{2c_1+c_2-1}$ are the challenger mutations. It remains to investigate whether these can be improving. The ESS–condition $\Pi(\beta \mid \alpha^0 \times \beta) < \Pi(\alpha^0 \mid \alpha^0 \times \alpha^0)$ is satisfied:

$$\text{iff } (1-c_1)^2 + 2(1-c_2)^2 - 2c_1(1-c_2) < 0, \text{ in case } \beta = \frac{c_2}{2c_2+c_1-1}, \tag{D.21}$$

and

$$\text{iff } 1 - \frac{1}{2}c_2 < c_1, \text{ in case } \beta = \frac{c_1}{2c_1+c_2-1}. \tag{D.22}$$

For all $(c_1, c_2) \in T_{2.a}$, D.21 implies D.22.

D.4 Technology 2.b

Lemma D.2 *For all technologies* $(c_1, c_2) \in T_{2.b}$*:*

$$\begin{array}{c}\alpha^2 = \frac{1}{c_1+c_2} \textit{ sustains } (\sigma^*_{11}, \sigma^*_{11}) \textit{ and is an ESS if} \\ 1 - c_2 - (2 - c_1 - c_2)^2 < 0,\end{array} \tag{D.23}$$

$$\begin{array}{c}\alpha^0 \textit{ sustains } (\sigma^*_{01}, \sigma^*_{01}) \textit{ and is an ESS if} \\ 1 - \frac{1}{2}c_2 < c_1.\end{array} \tag{D.24}$$

Proof. Since $c_1 \geq c_2$ and $c_2 \geq \frac{1}{2}$: $\alpha^2 \leq \alpha^1$ and therefore α^2 is the ESS-candidate supporting $(\sigma^*_{11}, \sigma^*_{11})$. Furthermore, since $\Delta S_1(\alpha, \alpha) > 0$ for all $\alpha > \alpha^2$ and 5.12: α^0 is the ESS-candidate supporting $(\sigma^*_{01}, \sigma^*_{01})$.

Candidate: α^2

By lemma 5.3 it suffices to consider mutations $\beta > \alpha^2$.

Feasible Plays, Match (α, β): Since $\Delta S_1(\alpha^2, \alpha^2) = 0$ and $\Delta S_1(\alpha, \beta)$ strictly increasing in β:

$$\Delta S_1(\alpha^2, \beta) > 0 \forall \beta \in \left(\alpha^2, 1\right]. \tag{D.25}$$

Furthermore,

$$\Delta U_1^1(\alpha^2, \beta = 1) = \frac{c_2 - c_1}{c_1 + c_2} \leq 0 \forall (c_1, c_2) \in T_{2.b},$$

and $\Delta U_1^1(\alpha, \beta)$ is increasing in β such that:

$$\Delta U_1^1(\alpha^2, \beta) \leq 0 \forall \beta \in \left[\alpha^2, 1\right]. \tag{D.26}$$

By D.25 and D.26 all mutations $\beta > \alpha^2$ induce play $\sigma^*_{01}(\alpha, \beta)$.

Feasible Plays, Match (β, α): Since $\Delta S_1(\alpha^2, \alpha^2) = 0$ and $\Delta S_1(\beta, \alpha)$ strictly increasing in β:

$$\Delta S_1(\beta, \alpha^2) > 0 \forall \beta \in \left(\alpha^2, 1\right]. \tag{D.27}$$

Note further:

$$\begin{array}{c}\Delta U_1^1(\alpha^2, \alpha^2) = \Delta U_1^2(\alpha^2, \alpha^2) \leq 0, \\ \Delta U_1^1(\beta = 1, \alpha^2) \leq 0 \forall (c_1, c_2) \in T_{2.b}\end{array}$$

and $\Delta U_1^1(\beta, \alpha)$ monotonic in β such that

$$\Delta U_1^1(\beta,\alpha) \leq 0 \forall \beta > \alpha^2. \tag{D.28}$$

By D.27 and D.28 all mutations $\beta > \alpha^2$ induce play $\sigma_{01}^*(\beta,\alpha)$.

Hence, all mutations $\beta > \alpha^2$ imply SPE production vector $(\sigma_{01}^*, \sigma_{01}^*)$. By lemma 5.3 the challenger mutation is $\beta = 1$. If

$$1 - c_2 - (2 - c_1 - c_2)^2 < 0 \Longleftrightarrow \Pi(\beta \mid \alpha^2 \times \beta) < \Pi(\alpha^2 \mid \alpha^2 \times \alpha^2)$$

the challenger mutation earns less than α^2 and the latter is an ESS. This proves D.23.

Candidate: α^0

Feasible Plays, Match (α,β): Firstly,

$$\Delta S_1(\alpha^0,\beta) \leq 0 \Longleftrightarrow \beta \leq \frac{c_1}{2c_1 + c_2 - 1} \tag{D.29}$$

$$\text{and } \Delta U_1^2(\alpha^0,\beta) \leq 0 \Longleftrightarrow \beta \leq \frac{1}{2c_2}. \tag{D.30}$$

Secondly,

$$\frac{c_1}{2c_1 + c_2 - 1} \leq \frac{1}{2c_2} \forall (c_1, c_2) \in T_{2.b} \tag{D.31}$$

such that $\beta \leq \frac{c_1}{2c_1+c_2-1}$ implies $\sigma_{11}^*(\alpha^0,\beta)$.

Thirdly, if $\beta > \frac{c_1}{2c_1+c_2-1}$: $\Delta S_1(\alpha^0,\beta) > 0$. Since $\Delta U_1^1(\alpha^0,\alpha^0) \leq 0$ and $\Delta U_1^1(\alpha,\beta)$ increasing in β, it follows:

$$\Delta U_1^1(\alpha^0,\beta) \leq 0 \forall \beta \leq 1. \tag{D.32}$$

Thus, play in match (α,β) is σ_{11}^* for mutations $\beta \leq \frac{c_1}{2c_1+c_2-1}$ and σ_{01}^* otherwise.

Feasible Plays, Match (β,α): Note that

$$\Delta S_1(\beta,\alpha^0) \leq 0 \Longleftrightarrow \beta \leq \frac{c_2}{2c_2 + c_1 - 1} \text{ and} \tag{D.33}$$

$$\Delta U_1^2(\beta,\alpha^0) \leq 0 \Longleftrightarrow \beta \leq \frac{c_2}{3c_2 - 1}. \tag{D.34}$$

The latter requires $c_2 > \frac{1}{3}$ which holds for $T_{2.b}$. Since $c_2 \leq c_1 \Longrightarrow \frac{c_2}{2c_2+c_1-1} \leq \frac{c_2}{3c_2-1}$, all mutations $\beta \leq \frac{c_2}{2c_2+c_1-1}$ lead to play $\sigma_{11}^*(\beta,\alpha^0)$.

If $\beta > \frac{c_2}{2c_2+c_1-1} \Longrightarrow \Delta S_1(\beta,\alpha^0) > 0$. Since $c_2 \geq c_1$ implies

$$\Delta U_1^1(\beta,\alpha^0) = \beta(c_2 - c_1) \leq 0 \forall \beta \in \left[\tfrac{1}{2}, 1\right], \tag{D.35}$$

it follows: Play in match (β, α) is σ^*_{11} for mutations $\beta \leq \frac{c_2}{2c_2+c_1-1}$ and σ^*_{01} otherwise.

For all $(c_1, c_2) \in T_{2.b}$:

$$\frac{c_2}{2c_2 + c_1 - 1} \leq \frac{c_1}{2c_1 + c_2 - 1}.$$

From here on the arguments are equivalent to those in D.1, statement D.20 and further.

D.5 Technology 2.c.1

Lemma D.3 *For all technologies* $(c_1, c_2) \in T_{2.c.1}$:

$$\alpha^2 = \tfrac{1}{c_1+c_2} \textit{ sustains } (\sigma^*_{11}, \sigma^*_{11}) \textit{ and is an ESS if } \quad 1 - c_2 - (2 - c_1 - c_2)^2 < 0, \tag{D.36}$$

$$\alpha^0 \textit{ sustains } (\sigma^*_{10}, \sigma^*_{10}) \textit{ and is not neutrally stable if } \quad 2 - 2c_1 > c_2. \tag{D.37}$$

Proof. By lemma 5.2 and the definition of $T_{2.c.1}$ it follows: α^2 is the ESS-candidate supporting $(\sigma^*_{11}, \sigma^*_{11})$ and α^0 implies $(\sigma^*_{10}, \sigma^*_{10})$.

Candidate: α^2

It suffices to consider mutations $\beta > \alpha^2$ (lemma 5.3).

Feasible Plays, Match (α, β): Since $\Delta S_1(\alpha^2, \alpha^2) = 0$ and $\Delta S_1(\alpha, \beta)$ strictly increasing in β:

$$\Delta S_1(\alpha^2, \beta) > 0 \forall \beta \in \left(\alpha^2, 1\right]. \tag{D.38}$$

Furthermore, note that

$$\Delta U_1^1(\alpha^2, \alpha^2) = \Delta U_1^2(\alpha^2, \alpha^2) \leq 0,$$
$$\Delta U_1^1(\alpha^2, \beta = 1) = \frac{c_2 - c_1}{c_1 + c_2} \geq 0 \forall (c_1, c_2) \in T_{2.c.1},$$

$$\Delta U_1^1(\alpha, \beta) \text{ monotonically increasing in } \beta$$

and

$$\Delta U_1^1(\alpha^2, \beta) = 0 \Longleftrightarrow \beta = \tilde{\beta} \equiv \frac{1 - c_2 - \alpha^2(2 - c_1 - c_2)}{1 - c_2 - \alpha^2(2 - c_1 - c_1)}.$$

Therefore:[1]

$$\Delta U_1^1(\alpha^2, \beta) \leq 0 \forall \beta \in \left[\alpha^2, \tilde{\beta}\right] \text{ and } \Delta\, U_1^1(\alpha^2, \beta) > 0 \forall \beta \in \left(\tilde{\beta}, 1\right]. \tag{D.39}$$

Thus in match (α, β) mutations $\alpha^2 < \beta \leq \tilde{\beta}$ induce play $\sigma^*_{01}(\alpha, \beta)$ and mutations $\tilde{\beta} < \beta \leq 1$ induce play $\sigma^*_{10}(\alpha, \beta)$.

[1] Substituting α^2 to get $\tilde{\beta}$ entirely in terms of the parameters c_1 and c_2 does not simplify the formula very much and it won't be necessary for the rest of the analysis.

Feasible Plays, Match (β, α): Notice $\Delta S_1(\alpha^2, \alpha^2) = 0$ and $\Delta S_1(\beta, \alpha)$ strictly increasing in β such that:

$$\Delta S_1(\beta, \alpha^2) > 0 \forall \beta \in \left(\alpha^2, 1\right]. \tag{D.40}$$

In addition:

$$\begin{aligned} \Delta U_1^1(\alpha^2, \alpha^2) &= \Delta U_1^2(\alpha^2, \alpha^2) \leq 0 \text{ and} \\ \Delta U_1^1(\beta = 1, \alpha^2) &\leq 0 \forall (c_1, c_2) \in T_{2.c.1}. \end{aligned} \tag{D.41}$$

To see the latter note, firstly, that $\Delta U_1^1(\beta, \alpha)$ is increasing in β and c_2 and decreasing in c_1. Secondly:

$$\Delta U_1^1(\beta = 1, \alpha^2) = \frac{\left[c_1^2 + c_1 c_2 - 3c_1 + 1\right]}{2 - c_1 - c_2}. \tag{D.42}$$

The sign of the right-hand expression in D.42 depends solely on the numerator [...], since the denominator is positive. So, the question is, whether [...] can become positive given the restrictions on c_1 and c_2. Since c_1 is bounded below, [...] is maximal at this bound. Evaluating [...] at $c_1 = \frac{1+c_2^2-c_2}{2-c_2}$ gives the following expression: $\frac{4c_2^2-4c_2+1}{3(c_2-2)}$, which is non–positive for all $c_2 \in \left[\frac{1}{2}, 1\right]$. Thus, D.41 holds. Accordingly, by D.40 and D.41 all mutations $\beta > \alpha^2$ induce play $\sigma_{01}^*(\beta, \alpha)$.

Taking together the results for both matches, the following SPE production vectors are feasible via mutations $\beta > \alpha^2$:

$$\begin{aligned} &(\sigma_{01}^*, \sigma_{01}^*) \quad \text{if } \alpha^2 < \beta \leq \tilde{\beta}, \\ &(\sigma_{10}^*, \sigma_{01}^*) \quad \text{if } \tilde{\beta} < \beta \leq 1. \end{aligned}$$

By lemma 5.1: $\tilde{\beta}$ and $\beta = 1$ are the challenger mutations. It remains to investigate whether these can be improving. The ESS-condition $\Pi(\beta \mid \alpha^2 \times \beta) < \Pi(\alpha^2 \mid \alpha^2 \times \alpha^2)$ is satisfied:

$$\text{iff } s_1^*(\alpha^2, \alpha^2) < 1 - c_1 + s_1^*(\alpha^2, \tilde{\beta}), \text{ in case } \beta = \tilde{\beta}, \tag{D.43}$$

and

$$\text{iff } s_1^*(\alpha^2, \alpha^2) < 2 - c_1 - c_2, \text{ in case } \beta = 1. \tag{D.44}$$

Since D.44 implies D.43, it suffices to examine D.44. The latter holds, if and only if

$$1 - c_2 - (2 - c_1 - c_2)^2 < 0,$$

which proves D.36.

Candidate: α^0

Feasible Plays, Match (α,β): Statements D.29, D.30 and D.31 (see lemma D.2) apply to technology $T_{2.c.1}$ as well such that $\beta \leq \frac{c_1}{2c_1+c_2-1}$ implies $\sigma^*_{11}(\alpha,\beta)$.

If $\beta > \frac{c_1}{2c_1+c_2-1}$: $\Delta S_1(\alpha^0,\beta) > 0$. Furthermore, notice that

$$\Delta U^1_1(\alpha^0,\alpha^0) \geq 0,$$

$$\Delta U^1_1(\alpha,\beta) \text{ increasing in } \beta,$$

$$\Delta U^1_1(\alpha^0,\beta) = 0 \Longleftrightarrow \beta = \frac{c_1-1}{2c_1-c_2-1}$$

and

$$\frac{c_1}{2c_1+c_2-1} \leq \frac{c_1-1}{2c_1-c_2-1} \forall (c_1,c_2) \in T_{2.c.1}.$$

Thus, play in match (α,β) is characterized by

$$\begin{array}{ll} \sigma^*_{11}(\alpha,\beta) & \text{for all } \beta \leq \frac{c_1}{2c_1+c_2-1}, \\ \sigma^*_{01}(\alpha,\beta) & \text{for all } \frac{c_1}{2c_1+c_2-1} < \beta \leq \frac{c_1-1}{2c_1-c_2-1} \text{ and} \\ \sigma^*_{10}(\alpha,\beta) & \text{otherwise.} \end{array} \tag{D.45}$$

Feasible Plays, Match (β,α): Since

$$\begin{aligned} \Delta S_1(\beta,\alpha^0) \leq 0 &\Longleftrightarrow \beta \leq \frac{c_2}{2c_2+c_1-1}, \\ \Delta U^2_1(\beta,\alpha^0) \leq 0 &\Longleftrightarrow \beta \leq \frac{c_2}{3c_2-1} \\ \frac{c_2}{3c_2-1} &\leq \frac{c_2}{2c_2+c_1-1} \forall (c_1,c_2) \in T_{2.c.1} \text{ and} \\ \Delta U^1_1(\beta,\alpha^0) &> 0 \forall (c_1,c_2) \in T_{2.c.1}, \end{aligned}$$

it follows: Play in match (β,α) is σ^*_{11} for all mutations $\beta \leq \frac{c_2}{3c_2-1}$ and σ^*_{10} otherwise.

Now, combining the results for both matches and recognizing that

$$\frac{c_2}{3c_2-1} \leq \frac{c_1}{2c_1+c_2-1} \forall (c_1,c_2) \in T_{2.c.1}$$

the following SPE production vectors are feasible via mutation:

$$\begin{array}{ll} (\sigma^*_{11},\sigma^*_{11}) & \text{if } \beta \leq \frac{c_2}{3c_2-1}, \\ (\sigma^*_{11},\sigma^*_{10}) & \text{if } \frac{c_2}{3c_2-1} < \beta \leq \frac{c_1}{2c_1+c_2-1}, \\ (\sigma^*_{01},\sigma^*_{10}) & \text{if } \frac{c_1}{2c_1+c_2-1} < \beta \leq \frac{c_1-1}{2c_1-c_2-1} \text{ and} \\ (\sigma^*_{01},\sigma^*_{10}) & \text{otherwise.} \end{array} \tag{D.46}$$

Mutations inducing $(\sigma_{10}^*, \sigma_{10}^*)$ cannot be improving (lemma 5.1). The challenger mutations are $\beta = \frac{c_2}{3c_2-1}$, $\beta = \frac{c_1}{2c_1+c_2-1}$ as well as $\beta = \frac{c_1-1}{2c_1-c_2-1}$. It remains to investigate whether they can be improving. If

$$\Pi(\beta \mid \alpha^0 \times \beta) > \Pi(\alpha^0 \mid \alpha^0 \times \alpha^0) \tag{D.47}$$

— i.e. the entrant earns a strictly higher payoff than the incumbant —, then α^0 is not neutrally evolutionarily stable. Condition D.47 is satisfied

$$\text{iff } \frac{3}{5} > c_2, \text{ in case } \beta = \frac{c_2}{3c_2 - 1}, \tag{D.48}$$

$$\text{iff } 2 - 2c_1 > c_2, \text{ in case } \beta = \frac{c_1}{2c_1 + c_2 - 1} \tag{D.49}$$

and

$$\text{iff } c_1 > c_2, \text{ in case } \beta = \frac{c_1 - 1}{2c_1 - c_2 - 1}. \tag{D.50}$$

If any of these conditions hold, the respective mutation earns strictly more than α^0 such that α^0 is not neutrally stable. For all $(c_1, c_2) \in T_{2.c.1}$ D.50 is impossible and D.49 implies D.48.

D.6 Technology 2.c.2

Lemma D.4 *For all technologies* $(c_1, c_2) \in T_{2.c.2}$*:*

$$\alpha^1 = \tfrac{2-c_2}{1+c_2} \textit{ sustains } (\sigma^*_{11}, \sigma^*_{11}) \textit{ and is an ESS if } c_1 < \tfrac{4}{3} - \tfrac{2}{3}c_2, \tag{D.51}$$

$$\alpha^0 \textit{ sustains } (\sigma^*_{10}, \sigma^*_{10}) \textit{ and is not neutrally stable if } 2 - 2c_1 > c_2. \tag{D.52}$$

Proof. By lemma 5.2 and the definition of $T_{2.c.2}$ it follows: α^1 is the ESS-candidate supporting $(\sigma^*_{11}, \sigma^*_{11})$ and α^0 implies $(\sigma^*_{10}, \sigma^*_{10})$.

Candidate: α^1

By lemma 5.3 it suffices to consider mutations $\beta > \alpha^1$.

Feasible Plays, Match (α, β): Firstly, note that

$$\Delta S_1(\alpha^1, \alpha^1) \leq 0,$$

$$\Delta S_1(\alpha, \beta) \text{ strictly increasing in } \beta,$$

$$\Delta S_1(\alpha, \beta) = 0 \Longleftrightarrow \beta = \dot{\beta} \equiv \frac{\alpha(1 + c_1) - 1}{2\alpha(c_1 + c_2) - (1 + c_2)}$$

and therefore

$$\Delta S_1(\alpha^1, \beta) \leq 0 \forall \beta \in \left[\alpha^1, \dot{\beta}\right]. \tag{D.53}$$

Secondly,

$$\Delta U_1^2(\alpha^1, \alpha^1) = 0,$$

and

$$\Delta U_1^2(\alpha, \beta) \text{ strictly increasing in } \beta$$

such that

$$\Delta U_1^2(\alpha^1, \beta) > 0 \forall \beta \in \left(\alpha^1, 1\right]. \tag{D.54}$$

If $\beta > \dot{\beta}$:

$$\Delta S_1(\alpha^1, \beta) > 0 \text{ and} \tag{D.55}$$

$$\Delta U_1^1(\alpha^1, \beta) > 0. \tag{D.56}$$

The latter can be shown by evaluating $\Delta U_1^1(\alpha^1, \beta)$ at $\dot{\beta}$ and $c_1 = \frac{1+c_2^2-c_2}{2-c_2}$, which results in $\Delta U_1^1(\alpha^1, \beta) = 0$. Since $c_1 = \frac{1+c_2^2-c_2}{2-c_2}$ is an upper bound for c_1 and furthermore $\Delta U_1^1(\alpha, \beta)$ is strictly decreasing in c_1 and increasing in β, D.56 follows. According to conditions D.53 to D.56 play in match (α, β) is $\sigma^*_{10}(\alpha, \beta)$ for all mutations $\alpha^1 < \beta$.

Feasible Plays, Match (β, α): Since

$$\Delta S_1(\alpha^1, \alpha^1) \leq 0,$$

$$\Delta S_1(\beta, \alpha^1) \text{ strictly increasing in } \beta$$

and

$$\Delta S_1(\beta, \alpha) = 0 \Longleftrightarrow \beta = \ddot{\beta} \equiv \frac{\alpha(1 + c_2) - 1}{2\alpha(c_1 + c_2) - (1 + c_1)}$$

it follows:

$$\Delta S_1(\beta, \alpha^1) \leq 0 \forall \beta \in \left[\alpha^1, \ddot{\beta}\right]. \tag{D.57}$$

Furthermore, notice that

$$\Delta U_1^2(\alpha^1, \alpha^1) = 0$$

and

$$\Delta U_1^2(\beta, \alpha) \text{ is strictly increasing in } \beta, \text{ since } \alpha^1 > \frac{3 - c_2}{2(1 + c_2)},$$

such that

$$\Delta U_1^2(\beta, \alpha^1) > 0 \forall \beta \in \left(\alpha^1, 1\right]. \tag{D.58}$$

Thus play in match (β, α) is σ^*_{10} $\forall \beta \in \left(\alpha^1, \ddot{\beta}\right]$. Together with the result for match (α, β), this implies that every such mutation induces SPE production vector $(\sigma^*_{10}, \sigma^*_{10})$. Before analyzing the effect of mutations $\beta > \ddot{\beta}$ one should notice that mutations inducing $(\sigma^*_{10}, \sigma^*_{10})$ can not be improving. This holds, since for all $c_2 \in T_{2.c.2}$:

$$\begin{aligned} \Pi(\beta \mid \alpha^1 \times \beta) &< \Pi(\alpha^1 \mid \alpha^1 \times \alpha^1) \\ &\Longleftrightarrow \\ 0 &< 1 - c_2 \end{aligned} \tag{D.59}$$

Now, back to the analysis of match (β, α). If $\beta > \ddot{\beta}$, then $\Delta S_1(\beta, \alpha^1) > 0$ and therefore play depends on $\Delta U_1^1(\beta, \alpha^1)$. If $\Delta U_1^1(\beta, \alpha^1) > 0$, this implies σ^*_{10}, which can not be improving — as shown above —, since the according production vector is $(\sigma^*_{10}, \sigma^*_{10})$.

Thus, it is only necessary to consider mutations β such that $\Delta U_1^1(\beta, \alpha^1) \leq 0$. Since $\Delta U_1^1(\beta, \alpha^1)$ is decreasing in β for all $(c_1, c_2) \in$

$T_{2.c.2}$, it follows: If there exists $\beta \in \left(\ddot{\beta}, 1\right]$ such that $\Delta U_1^1(\beta, \alpha^1) \leq 0$, then the latter holds also for $\beta = 1$.

By this argument and lemma 5.1 it suffices to investigate whether $\beta = 1$ implies $\Delta S_1(\beta, \alpha^1) > 0$ and $\Delta U_1^1(\beta, \alpha^1) \leq 0$ and whether the according production vector $(\sigma_{10}^*, \sigma_{01}^*)$ can be improving. One can easily see that

$$\Delta S_1(\beta = 1, \alpha) > 0 \Longleftrightarrow [\alpha(2c_1 + c_2 - 1) - c_1] > 0.$$

Evaluating [...] at α^1results in

$$\frac{(1 - c_2)(c_2 + 3c_1 - 2)}{1 + c_2} > 0,$$

which holds for all $(c_1, c_2) \in T_{2.c.2}$. Since

$$\begin{aligned} \Delta U_1^1(\beta = 1, \alpha^1) &\leq 0 \\ &\Longleftrightarrow \\ \frac{1}{3}(1 + c_2) &\leq c_1, \end{aligned} \tag{D.60}$$

the latter condition implies play σ_{01}^* in match (β, α) and thus production vector $(\sigma_{10}^*, \sigma_{01}^*)$. If D.60 is satisfied, $\beta = 1$ might eventually be improving, but this remains to be checked. Given mutation $\beta = 1$ and $(\sigma_{10}^*, \sigma_{01}^*)$ — i.e. D.60 holds:

$$\begin{aligned} \Pi(\beta \mid \alpha^1 \times \beta) &\geq \Pi(\alpha^1 \mid \alpha^1 \times \alpha^1) \\ &\Longleftrightarrow \\ c_1 &\geq \frac{4}{3} - \frac{2}{3}c_2. \end{aligned} \tag{D.61}$$

Since for all $(c_1, c_2) \in T_{2.c.2}$ condition D.61 implies condition D.60 this proves D.51.

Candidate: α^0

All arguments which were given to prove D.37 in lemma D.3 hold for technology $T_{2.c.2}$ as well.

D.7 Technology 2.d

Lemma D.5 *For all technologies* $(c_1, c_2) \in T_{2.d}$*:*

$$\alpha^1 = \tfrac{2-c_2}{1+c_2} \text{ sustains } (\sigma_{11}^*, \sigma_{11}^*) \text{ and is an ESS,} \tag{D.62}$$

$$\alpha^0 \text{ sustains } (\sigma_{10}^*, \sigma_{10}^*) \text{ and is not neutrally stable.} \tag{D.63}$$

Proof. By 5.2 and the definition of $T_{2.d}$: α^1 is the ESS-candidate supporting $(\sigma_{11}^*, \sigma_{11}^*)$ and α^0 implies $(\sigma_{10}^*, \sigma_{10}^*)$.

Candidate: α^1

All arguments up to condition D.59 which were given to prove D.51 in lemma D.4 hold for technology $T_{2.d}$ as well. Therefore the analysis proceeds by examining whether in match (β, α) mutations $\beta > \ddot{\beta}$ can lead to play $\sigma_{01}^*(\beta, \alpha)$. Recall that

$$\ddot{\beta} \equiv \frac{\alpha(1+c_2)-1}{2\alpha(c_1+c_2)-(1+c_1)}$$

and that $\beta > \ddot{\beta}$ implies $\Delta S_1(\beta, \alpha^1) > 0$ such that play in (β, α) depends on $\Delta U_1^1(\beta, \alpha^1)$. It will be shown that the latter is always positive such that no impoving mutation exists. To see this notice first: c_1 is bounded above by $\frac{1}{2}$ given $T_{2.d}$, and

$$\Delta U_1^1(\ddot{\beta}, \alpha^1) \mid_{c_1=\frac{1}{2}} > 0 \forall (c_1, c_2) \in T_{2.d},$$

as well as

$$\Delta U_1^1(\beta, \alpha) \text{ is decreasing in } c_1,$$

such that

$$\Delta U_1^1(\ddot{\beta}, \alpha^1) > 0 \forall (c_1, c_2) \in T_{2.d}. \tag{D.64}$$

Secondly, if

$$c_1 \leq \frac{2}{3} - \frac{1}{3}c_2 \tag{D.65}$$

then

$$\Delta U_1^1(\beta, \alpha^1) \text{ is increasing in } \beta$$

and therefore

$$\Delta U_1^1(\beta, \alpha^1) \mid_{c_1 \leq \frac{2}{3}-\frac{1}{3}c_2} > 0 \tag{D.66}$$

$$\forall \beta \in \left[\ddot{\beta}, 1\right] \wedge (c_1, c_2) \in T_{2.d}.$$

Thirdly, if

$$c_1 > \frac{2}{3} - \frac{1}{3}c_2 \tag{D.67}$$

then

$$\Delta U_1^1(\beta, \alpha^1) \text{ is decreasing in } \beta.$$

Finally, evaluating $\Delta U_1^1(\beta, \alpha^1)$ at $\beta = 1$ one gets:

$$\Delta U_1^1(\beta = 1, \alpha^1) = \frac{1}{3}(1 + c_2 - 3c_1) > 0 \forall (c_1, c_2) \in T_{2.d},$$

and hence

$$\Delta U_1^1(\beta, \alpha^1) \mid_{c_1 > \frac{2}{3} - \frac{1}{3}c_2} > 0 \tag{D.68}$$

$$\forall \beta \in \left[\ddot{\beta}, 1\right] \wedge (c_1, c_2) \in T_{2.d}.$$

Thus, in match (β, α) all mutations $\beta > \alpha^1$ induce play σ_{10}^*. This implies that all such mutations lead to production vector $(\sigma_{10}^*, \sigma_{10}^*)$ and are not improving, as shown above within the proof of lemma D.4. This proves D.62.

Candidate: α^0

It suffices to give an example for a mutation that is feasible and improving for all $(c_1, c_2) \in T_{2.d}$. Such a mutation is $\beta = \frac{1}{2c_2}$. It implies $\Delta U_1^2(\alpha^0, \beta) = 0$. Since

$$\Delta S_1(\alpha^0, \beta) \leq 0 \Longleftrightarrow \beta \leq \frac{c_1}{2c_1 + c_2 - 1}$$

and

$$\frac{1}{2c_2} \leq \frac{c_1}{2c_1 + c_2 - 1} \forall (c_1, c_2) \in T_{2.d}$$

it follows: $\beta = \frac{1}{2c_2}$ leads to play σ_{11}^* in match (α, β).
In match (β, α):

$$\Delta S_1(\beta, \alpha^0) \leq 0 \Longleftrightarrow \beta \leq \frac{c_2}{2c_2 + c_1 - 1} \text{ and}$$

$$\Delta U_1^2(\beta, \alpha^0) \leq 0 \Longleftrightarrow \beta \leq \frac{c_2}{3c_2 - 1}.$$

Since for all $(c_1, c_2) \in T_{2.d}$

$$\frac{c_2}{3c_2 - 1} \leq \frac{1}{2c_2} \leq \frac{c_2}{2c_2 + c_1 - 1},$$

it follows: $\beta = \frac{1}{2c_2}$ leads to play σ^*_{10} in match (β, α). Thus, mutation $\beta = \frac{1}{2c_2}$ induces SPE production vector $(\sigma^*_{11}(\alpha, \beta), \sigma^*_{10}(\beta, \alpha))$. Since

$$\begin{aligned} \Pi(\beta \mid \alpha^0 \times \beta) &> \Pi(\alpha^0 \mid \alpha^0 \times \alpha^0) \\ &\Longleftrightarrow \\ 1 - c_2 &> 0 \end{aligned}$$

holds for all $(c_1, c_2) \in T_{2.d}$ this concludes the proof.

References

1. Albers, W. (1997), Foundations of a theory of prominence in the decimal system — Part I: Numerical response as a process, exactness, scales, and structure of scales, *Discussion Paper*, Institute of Mathematical Economics, No. 265, University of Bielefeld.
2. Albers, W. and G. Albers (1983), On the prominence structure of the decimal system, in: R. W. Scholz (ed.), *Decision Making under Uncertainty*, Amsterdam: Elsevier Science Publishers, 271-287.
3. Albers, W., E. Albers, L. Albers and B. Vogt (1997), Foundations of a theory of prominence in the decimal system — Part II: Exactness selection rule, and confirming results, *Discussion Paper*, Institute of Mathematical Economics, No. 266, University of Bielefeld.
4. Alchian, A. A. (1950), Uncertainty, evolution, and economic theory, *Journal of Political Economy*, Vol. 58, 211-221.
5. Axelrod, R. (1986), An evolutionary approach to norms, *American Political Science Review*, Vol. 80, 1095-1111.
6. Becker, G. S. (1976), Altruism, egoism, and genetic fitness: Economics and sociobiology, *Journal of Economic Literature*, Vol. 14, 817-826.
7. Bester, H. and W. Güth (1998), Is altruism evolutionarily stable?, *Journal of Economic Behavior and Organization,* Vol. 34 (2), 193-209.
8. Berg, J., J. Dickhaut and K. A. McCabe (1995), Trust, reciprocity and social history, *Games and Economic Behavior*, Vol. 10, 122-142.
9. Binmore, K. and L. Samuelson (1994), An economist's perspective on the evolution of of norms: Comment, *Journal of Institutional and Theoretical Economics*, Vol. 150, 68-71.
10. Binmore, K.G., A. Shaked and J. Sutton (1985): Testing noncooperative bargaining theory: A preliminary study, *American Economic Review*, Vol. 78, 1178-1180.
11. Binmore, K., J. Swierzbinski, S. Hsu and C. Proulx (1993), Focal points and bargaining, *International Journal of Game Theory*, Vol. 22, 381-409.
12. Bolton, G.E. (1991): A comparative model of bargaining: Theory and evidence, *American Economic Review*, Vol. 81, 1096-1136.
13. Bolton, G.E. and R. Zwick (1995): Anonymity versus punishment in ultimatum bargaining, *Games and Economic Behavior*, Vol. 10, 95-121.
14. Brandstätter, H. (1988), Sechzehn Persönlichkeits–Adjektivskalen (16 PA) als Forschungsinstrument anstelle des 16 PF, *Zeitschrift für experimentelle und angewandte Psychologie*, Band 35, 370-391.
15. Brandstätter, H. and M. Königstein (1998), Personality influences on ultimatum bargaining decisions, *Working Paper*, Humboldt–University at Berlin.
16. Bortz, J. (1993), *Statistik für Sozialwissenschaftler*, 4. edition, Berlin: Springer Verlag.

17. Buchanan, J. M. (1975), *The limits of liberty*, Chicago: University of Chicago Press.
18. Camerer, C. and R. H. Thaler (1995), Anomalies — Ultimatums, dictators and manners, *Journal of Economic Perspectives*, Vol. 9, 209-219.
19. Davis, D. D. and C. A. Holt (1993), *Experimental Economics*, Princeton: Princeton University Press.
20. Dufwenberg, M. and W. Güth (1999), Indirect evolution versus strategic delegation: A comparison of two approaches to explaining economic institutions, *European Journal of Political Economy*, Vol. 15, 281-295.
21. Fehr, E., S. Gächter and G. Kirchsteiger (1997), Reciprocity as a contract enforcement device — Experimental evidence, *Econometrica*, Vol. 65 (4), 833-860.
22. Forsythe, R., J. L. Horowitz, N. E. Savin and M. Sefton (1994), Fairness in simple bargaining experiments, *Games and Economic Behavior*, Vol. 6, 347–369.
23. Frey, B. S. (1990), *Ökonomie ist Sozialwissenschaft — Die Anwendung der Ökonomie auf neue Gebiete*, München: Vahlen.
24. Gantner, A., W. Güth and M. Königstein (1998), Equity anchoring in simple bargaining games with production, *SFB-Discussion Paper*, No. 91-1998, Humboldt–University at Berlin.
25. Gauthier, D. (1986), *Morals by agreement*, Oxford: Oxford University Press.
26. Greene, W. and T. Seaks (1991), The restricted least squares estimator: A pedagogical note, *The Review of Economics and Statistics*, 563-567.
27. Güth, W. (1988), On the behavioral approach to distributive justice — A theoretical and experimental investigation, in: S. Maital, (ed.), *Applied Behavioral Economics*, Vol. II., Brighton/Sussex: Wheatsheaf Books, 703-717.
28. Güth, W. (1994): Distributive justice — A behavioral theory and empirical evidence, in: H. Brandstätter and W. Güth (eds.), *Essays on Economic Psychology*, Berlin: Springer, 153-176.
29. Güth, W. (1995a), An evolutionary approach to explaining cooperative behavior by reciprocal incentives, *International Journal of Game Theory*, Vol. 24, 323-344.
30. Güth, W. (1995b), On ultimatum bargaining experiments — A personal review, *Journal of Economic Behavior and Organization*, Vol. 27, 329-344.
31. Güth, W. and S. Huck (1997), From ultimatum bargaining to dictatorship — An experimental study of four games varying in veto power, *Metroeconomica*, Vol. 48 (3), 262-279.
32. Güth, W., S. Huck and P. Ockenfels (1996), Two-level-ultimatum bargaining with incomplete information — An experimental study, *The Economic Journal*, Vol. 106, No. 436, 593-604.
33. Güth, W., S. Huck and A. Rapoport (1998), The limitations of the positional order effect, *Journal of Economic Behavior and Organization*, Vol. 34 (2), 313-325.
34. Güth, W. and H. Kliemt (1994), Competition or co–operation, *Metroeconomica*, Vol. 45, 155-187.
35. Güth, W., W. Klose, M. Königstein and J. Schwalbach (1998), An experimental study of a dynamic principal–agent relationship, *Managerial and Decision Economics*, Vol. 19, 327-341.
36. Güth, W. and B. Peleg (1997), On the survival of the fittest — An indirect evolutionary approach based on differentiability, *Working Paper*, Humboldt–University at Berlin.

37. Güth, W., R. Schmittberger and B. Schwarze (1982), An experimental analysis of ultimatum bargaining, *Journal of Economic Behavior and Organization*, Vol. 3, 367-388.
38. Güth, W. and R. Tietz (1986): Auctioning ultimatum bargaining positions — How to act if rational decisions are unacceptable, in: R.W. Scholz (ed.), *Current Issues in West German Decision Research*, Frankfurt, 173-185.
39. Güth, W. and R. Tietz (1990), Ultimatum bargaining behavior — A survey and comparison of experimental results, *Journal of Economic Psychology*, Vol. 11, 417-449.
40. Güth, W. and M. Yaari (1992), An evolutionary approach to explain reciprocal behavior in a simple strategic game, in: U. Witt (ed.), *Explaining Process and Change — Approaches to Evolutionary Economics*, Ann Arbor, 23-34.
41. Hackett, S. (1993), An experimental analysis of surplus sharing in incomplete contracts, *Discussion Paper*, Indiana University, Bloomington.
42. Hammerstein, P. and R. Selten (1994), Game theory and evolutionary biology, in: R. Aumann and S. Hart (eds.), *Handbook of Game Theory*, Vol. 2, 929-993.
43. Harrison, G. W. and J. Hirshleifer (1989), An experimental evaluation of weakest link/best shot models of public goods, *Journal of Political Economy*, Vol. 97, 201-225.
44. Harrison, G. W. and K. A. McCabe (1992), Expectations and fairness in a simple bargaining experiment, *Working Paper*, B-92-10, University of South Carolina.
45. Hirshleifer, J. (1977), Economics from a biological viewpoint, *Journal of Law and Economics*, Vol. 20 (1), 1-52.
46. Hobbes, T. (1651), *Leviathan*, 1991 edition, edited by R. Tuck, Cambridge: Cambridge University Press.
47. Hoffman, E., K. McCabe, K. Shachat, and V.A. Smith (1994): Preferences, property rights, and anonymity in bargaining games, *Games and Economic Behavior*, Vol. 7, 346-380.
48. Hoffman, E. and M. L. Spitzer (1985): Entitlements, rights, and fairness: An experimental examination of subjects' concepts of distributive justice, *Journal of Legal Studies*, Vol. 14, 259-297.
49. Homans, G. C. (1961), *Social Behavior: Its Elementary Forms*, London.
50. Huck, S. (1998), Trust, treason, and trials: An example of how the evolution of preferences can be driven by legal institutions, *Journal of Law, Economics, & Organization*, Vol. 14, 44-60.
51. Huck, S. and J. Oechsler (1999), The indirect evolutionary approach to explaining fair allocations, *Games and Economic Behavior*, Vol. 28, 13-24.
52. Judge, G. G, W. E. Griffiths, R. Carter Hill, H. Lütkepohl and T. Lee (1985), *The Theory and Practice of Econometrics*, 2. edition, New York: John Wiley and Sons.
53. Kerlinger, F. N. and E. J. Pedhazur (1973), *Multiple Regression in Behavioral Research*, New York: Holt, Rinehart and Winston.
54. Königstein, M. (1996a), Rationalitätseinschränkungen im Teamwork–Experiment, *Working Paper*, Humboldt–University at Berlin.
55. Königstein, M. (1996b), Der Einsatz von Auktion und Planungsbericht zur Vorbereitung experimenteller Entscheidungen — Merkmale, theoretische Auswirkungen und empirische Evidenz, *Working Paper*, Humboldt–University at Berlin.
56. Königstein, M. (1998a), Convergence to equitable play in the repeated ultimatum game with advance production, *SFB-Discussion Paper*, No. 89-1998, Humboldt–University at Berlin.

57. Königstein, M. (1998b), Measuring treatment effects in experimental cross–sectional time series, *SFB-Discussion Paper*, No. 88-1998, Humboldt–University at Berlin.
58. Königstein, M. (1998c), Efficiency and evolution of social preferences and prosocial behavior, *SFB-Discussion Paper*, No. 90-1998, Humboldt–University at Berlin.
59. Königstein, M. and W. Müller (1998), Combining rational choice and evolutionary dynamics: The indirect evolutionary approach, *Discussion Paper, Economics Series*, No. 136, Humboldt–University at Berlin.
60. Königstein, M. and W. Müller (forthcoming), Combining rational choice and evolutionary dynamics: The indirect evolutionary approach, *Metroeconomica*.
61. Königstein, M. and R. Tietz (1998), Profit sharing in an asymmetric bargaining game, *SFB-Discussion Paper*, No. 87-1998, Humboldt–University at Berlin.
62. Loubergé, H. and H. Schlesinger (1988), Cutting the cake with a stranger — Egoism and altruism with Imperfect Information, *Journal of Economic Behavior and Organization*, Vol. 9, 377-388.
63. Lück, H. E. (1975), *Prosoziales Verhalten — Empirische Untersuchungen zur Hilfeleistung*, Köln: Kiepenheuer & Witsch.
64. Maynard Smith, J. (1982), *Evolution and the Theory of Games*, Cambridge.
65. Mikula, G. (1972), Gewinnaufteilungsverfahren in Dyaden bei variiertem Leistungsverhältnis, *Zeitschrift für Sozialpsychologie*, Vol. 3, 126-133.
66. Mitzkewitz, M. and R. Nagel (1993), Experimental Results on Ultimatum Games with Incomplete Information, *International Journal of Game Theory*, Vol. 22, 171-198.
67. Nash, J. F. (1950), The bargaining problem, *Econometrica*, Vol. 18, 155-162.
68. Prasnikar, V. and A. E. Roth (1992), Considerations of fairness and strategy: Experimental data from sequential games, *Quarterly Journal of Economics*, August, 865-888.
69. Rapoport, A., E. Weg and D. S. Felsenthal (1990), Effects of fixed costs in two–person sequential bargaining, *Theory and Decision*, Vol. 28, 47-72.
70. Roth, A. E. (1995), Bargaining experiments, in: J. Kagel and A. E. Roth (eds.), *Handbook of Experimental Economics*, 253-348.
71. Roth, A. E., V. Prasnikar, M. Okuno–Fujiwara and S. Zamir (1991), Bargaining and market behavior in Jerusalem, Ljubljana, Pittsburgh, and Tokyo: An experimental study. *American Economic Review*, Vol. 81, 1068-1095.
72. Selten, R. (1965), Spieltheoretische Behandlung eines Oligopolmodells mit Nachfrageträgheit, *Zeitschrift für die gesamte Staatswissenschaft*, Bd.121, Teil I, 301-324, Teil II, 667-689.
73. Selten, R.(1975), Reexamination of the Perfectness Concept for Equilibrium Points in Extensive Games, *International Journal of Game Theory*, Vol. 4, 25-55.
74. Selten, R. (1978), The equity principle in economic behavior, in: H. Gottinger and W. Leinfellner (Eds.), *Decision Theory and Social Ethics, Issues in Social Choice*, Dordrecht: Reidel, 289-301.
75. Selten, R. (1988), Evolutionary stability in extensive two–person games: Corrections and further development, *Mathematical Social Sciences*, Vol. 16, 223-266.
76. Selten, R. (1989), Properties of a measure of predictive success, *Discussion Paper* No. B–130, University of Bonn.
77. Selten, R. (1991), Properties of a measure of predictive success, *Mathematical Social Sciences*, Vol. 21, 153-167.
78. Selten, R. (1991), Evolution, learning and economic behavior, *Games and Economic Behavior*, Vol. 3, 3-24.

79. Selten, R. and W. Krischker (1983): Comparison of two theories for characteristic function experiments, in: R. Tietz (ed.), Aspiration levels in bargaining and economic decision making, Berlin: Springer, 259-264.
80. Sugden, R. (1993), The contractarian enterprise, in: D. Gauthier and R. Sugden (eds.), *Rationality, justice and the social contract — Themes form 'Morals by aggreement'*, Hertfordshire: Harvester Wheatsheaf, 1-23.
81. Suits, D. (1984), Dummy variables: Mechanics v. interpretation, *The Review of Economics and Statistics*, Vol. 66, 177-180.
82. Thaler, R. H. (1988), The ultimatum game, *Journal of Economic Perspectives*, Vol. 2, 195-206.
83. Vega-Redondo, F. (1996), *Evolution, Games, and Economic Behavior*, Oxford University Press.
84. Weibull, J. W. (1995), *Evolutionary Game Theory*, Cambridge/Mass.
85. Weg, E. and R. Zwick (1991), On the robustness of perfect equilibrium in fixed cost sequential bargaining under an isomorphic transformation, *Economics Letters*, Vol. 36, 21-24.

Lecture Notes in Economics and Mathematical Systems

For information about Vols. 1–295 please contact your bookseller or Springer-Verlag

Vol. 296: A. Börsch-Supan, Econometric Analysis of Discrete Choice. VIII, 211 pages. 1987.

Vol. 297: V. Fedorov, H. Läuter (Eds.), Model-Oriented Data Analysis. Proceedings, 1987. VI, 239 pages. 1988.

Vol. 298: S.H. Chew, Q. Zheng, Integral Global Optimization. VII, 179 pages. 1988.

Vol. 299: K. Marti, Descent Directions and Efficient Solutions in Discretely Distributed Stochastic Programs. XIV, 178 pages. 1988.

Vol. 300: U. Derigs, Programming in Networks and Graphs. XI, 315 pages. 1988.

Vol. 301: J. Kacprzyk, M. Roubens (Eds.), Non-Conventional Preference Relations in Decision Making. VII, 155 pages. 1988.

Vol. 302: H.A. Eiselt, G. Pederzoli (Eds.), Advances in Optimization and Control. Proceedings, 1986. VIII, 372 pages. 1988.

Vol. 303: F.X. Diebold, Empirical Modeling of Exchange Rate Dynamics. VII, 143 pages. 1988.

Vol. 304: A. Kurzhanski, K. Neumann, D. Pallaschke (Eds.), Optimization, Parallel Processing and Applications. Proceedings, 1987. VI, 292 pages. 1988.

Vol. 305: G.-J.C.Th. van Schijndel, Dynamic Firm and Investor Behaviour under Progressive Personal Taxation. X, 215 pages.1988.

Vol. 306: Ch. Klein, A Static Microeconomic Model of Pure Competition. VIII, 139 pages. 1988.

Vol. 307: T.K. Dijkstra (Ed.), On Model Uncertainty and its Statistical Implications. VII, 138 pages. 1988.

Vol. 308: J.R. Daduna, A. Wren (Eds.), Computer-Aided Transit Scheduling. VIII, 339 pages. 1988.

Vol. 309: G. Ricci, K. Velupillai (Eds.), Growth Cycles and Multisectoral Economics: The Goodwin Tradition. III, 126 pages. 1988.

Vol. 310: J. Kacprzyk, M. Fedrizzi (Eds.), Combining Fuzzy Imprecision with Probabilistic Uncertainty in Decision Making. IX, 399 pages. 1988.

Vol. 311: R. Färe, Fundamentals of Production Theory. IX, 163 pages. 1988.

Vol. 312: J. Krishnakumar, Estimation of Simultaneous Equation Models with Error Components Structure. X, 357 pages. 1988.

Vol. 313: W. Jammernegg, Sequential Binary Investment Decisions. VI, 156 pages. 1988.

Vol. 314: R. Tietz, W. Albers, R. Selten (Eds.), Bounded Rational Behavior in Experimental Games and Markets. VI, 368 pages. 1988.

Vol. 315: I. Orishimo, G.J.D. Hewings, P. Nijkamp (Eds.), Information Technology: Social and Spatial Perspectives. Proceedings 1986. VI, 268 pages. 1988.

Vol. 316: R.L. Basmann, D.J. Slottje, K. Hayes, J.D. Johnson, D.J. Molina, The Generalized Fechner-Thurstone Direct Utility Function and Some of its Uses. VIII, 159 pages. 1988.

Vol. 317: L. Bianco, A. La Bella (Eds.), Freight Transport Planning and Logistics. Proceedings, 1987. X, 568 pages. 1988.

Vol. 318: T. Doup, Simplicial Algorithms on the Simplotope. VIII, 262 pages. 1988.

Vol. 319: D.T. Luc, Theory of Vector Optimization. VIII, 173 pages. 1989.

Vol. 320: D. van der Wijst, Financial Structure in Small Business. VII, 181 pages. 1989.

Vol. 321: M. Di Matteo, R.M. Goodwin, A. Vercelli (Eds.), Technological and Social Factors in Long Term Fluctuations. Proceedings. IX, 442 pages. 1989.

Vol. 322: T. Kollintzas (Ed.), The Rational Expectations Equilibrium Inventory Model. XI, 269 pages. 1989.

Vol. 323: M.B.M. de Koster, Capacity Oriented Analysis and Design of Production Systems. XII, 245 pages. 1989.

Vol. 324: I.M. Bomze, B.M. Pötscher, Game Theoretical Foundations of Evolutionary Stability. VI, 145 pages. 1989.

Vol. 325: P. Ferri, E. Greenberg, The Labor Market and Business Cycle Theories. X, 183 pages. 1989.

Vol. 326: Ch. Sauer, Alternative Theories of Output, Unemployment, and Inflation in Germany: 1960–1985. XIII, 206 pages. 1989.

Vol. 327: M. Tawada, Production Structure and International Trade. V, 132 pages. 1989.

Vol. 328: W. Güth, B. Kalkofen, Unique Solutions for Strategic Games. VII, 200 pages. 1989.

Vol. 329: G. Tillmann, Equity, Incentives, and Taxation. VI, 132 pages. 1989.

Vol. 330: P.M. Kort, Optimal Dynamic Investment Policies of a Value Maximizing Firm. VII, 185 pages. 1989.

Vol. 331: A. Lewandowski, A.P. Wierzbicki (Eds.), Aspiration Based Decision Support Systems. X, 400 pages. 1989.

Vol. 332: T.R. Gulledge, Jr., L.A. Litteral (Eds.), Cost Analysis Applications of Economics and Operations Research. Proceedings. VII, 422 pages. 1989.

Vol. 333: N. Dellaert, Production to Order. VII, 158 pages. 1989.

Vol. 334: H.-W. Lorenz, Nonlinear Dynamical Economics and Chaotic Motion. XI, 248 pages. 1989.

Vol. 335: A.G. Lockett, G. Islei (Eds.), Improving Decision Making in Organisations. Proceedings. IX, 606 pages. 1989.

Vol. 336: T. Puu, Nonlinear Economic Dynamics. VII, 119 pages. 1989.

Vol. 337: A. Lewandowski, I. Stanchev (Eds.), Methodology and Software for Interactive Decision Support. VIII, 309 pages. 1989.

Vol. 338: J.K. Ho, R.P. Sundarraj, DECOMP: An Implementation of Dantzig-Wolfe Decomposition for Linear Programming. VI, 206 pages.

Vol. 339: J. Terceiro Lomba, Estimation of Dynamic Econometric Models with Errors in Variables. VIII, 116 pages. 1990.

Vol. 340: T. Vasko, R. Ayres, L. Fontvieille (Eds.), Life Cycles and Long Waves. XIV, 293 pages. 1990.

Vol. 341: G.R. Uhlich, Descriptive Theories of Bargaining. IX, 165 pages. 1990.

Vol. 342: K. Okuguchi, F. Szidarovszky, The Theory of Oligopoly with Multi-Product Firms. V, 167 pages. 1990.

Vol. 343: C. Chiarella, The Elements of a Nonlinear Theory of Economic Dynamics. IX, 149 pages. 1990.

Vol. 344: K. Neumann, Stochastic Project Networks. XI, 237 pages. 1990.

Vol. 345: A. Cambini, E. Castagnoli, L. Martein, P Mazzoleni, S. Schaible (Eds.), Generalized Convexity and Fractional Programming with Economic Applications. Proceedings, 1988. VII, 361 pages. 1990.

Vol. 346: R. von Randow (Ed.), Integer Programming and Related Areas. A Classified Bibliography 1984–1987. XIII, 514 pages. 1990.

Vol. 347: D. Ríos Insua, Sensitivity Analysis in Multiobjective Decision Making. XI, 193 pages. 1990.

Vol. 348: H. Störmer, Binary Functions and their Applications. VIII, 151 pages. 1990.

Vol. 349: G.A. Pfann, Dynamic Modelling of Stochastic Demand for Manufacturing Employment. VI, 158 pages. 1990.

Vol. 350: W.-B. Zhang, Economic Dynamics. X, 232 pages. 1990.

Vol. 351: A. Lewandowski, V. Volkovich (Eds.), Multiobjective Problems of Mathematical Programming. Proceedings, 1988. VII, 315 pages. 1991.

Vol. 352: O. van Hilten, Optimal Firm Behaviour in the Context of Technological Progress and a Business Cycle. XII, 229 pages. 1991.

Vol. 353: G. Ricci (Ed.), Decision Processes in Economics. Proceedings, 1989. III, 209 pages 1991.

Vol. 354: M. Ivaldi, A Structural Analysis of Expectation Formation. XII, 230 pages. 1991.

Vol. 355: M. Salomon. Deterministic Lotsizing Models for Production Planning. VII, 158 pages. 1991.

Vol. 356: P. Korhonen, A. Lewandowski, J . Wallenius (Eds.), Multiple Criteria Decision Support. Proceedings, 1989. XII, 393 pages. 1991.

Vol. 357: P. Zörnig, Degeneracy Graphs and Simplex Cycling. XV, 194 pages. 1991.

Vol. 358: P. Knottnerus. Linear Models with Correlated Disturbances. VIII, 196 pages. 1991.

Vol. 359: E. de Jong, Exchange Rate Determination and Optimal Economic Policy Under Various Exchange Rate Regimes. VII, 270 pages. 1991.

Vol. 360: P. Stalder, Regime Translations, Spillovers and Buffer Stocks. VI, 193 pages . 1991.

Vol. 361: C. F. Daganzo. Logistics Systems Analysis. X, 321 pages. 1991.

Vol. 362: F. Gehrels, Essays in Macroeconomics of an Open Economy. VII, 183 pages. 1991.

Vol. 363: C. Puppe, Distorted Probabilities and Choice under Risk. VIII, 100 pages . 1991

Vol. 364: B. Horvath, Are Policy Variables Exogenous? XII, 162 pages. 1991.

Vol. 365: G. A. Heuer, U. Leopold-Wildburger. Balanced Silverman Games on General Discrete Sets. V, 140 pages. 1991.

Vol. 366: J. Gruber (Ed.). Econometric Decision Models. Proceedings, 1989. VIII, 636 pages. 1991.

Vol. 367: M. Grauer, D. B. Pressmar (Eds.). Parallel Computing and Mathematical Optimization. Proceedings. V, 208 pages. 1991.

Vol. 368: M. Fedrizzi, J. Kacprzyk, M. Roubens (Eds.). Interactive Fuzzy Optimization. VII, 216 pages. 1991.

Vol. 369: R. Koblo, The Visible Hand. VIII, 131 pages.1991.

Vol. 370: M. J. Beckmann, M. N. Gopalan, R. Subramanian (Eds.), Stochastic Processes and their Applications. Proceedings, 1990. XLI, 292 pages. 1991.

Vol. 371: A. Schmutzler, Flexibility and Adjustment to Information in Sequential Decision Problems. VIII, 198 pages. 1991.

Vol. 372: J. Esteban, The Social Viability of Money. X, 202 pages. 1991.

Vol. 373: A. Billot, Economic Theory of Fuzzy Equilibria. XIII, 164 pages. 1992.

Vol. 374: G. Pflug, U. Dieter (Eds.), Simulation and Optimization. Proceedings, 1990. X, 162 pages. 1992.

Vol. 375: S.-J. Chen, Ch.-L. Hwang, Fuzzy Multiple Attribute Decision Making. XII, 536 pages. 1992.

Vol. 376: K.-H. Jöckel, G. Rothe, W. Sendler (Eds.), Bootstrapping and Related Techniques. Proceedings, 1990. VIII, 247 pages. 1992.

Vol. 377: A. Villar, Operator Theorems with Applications to Distributive Problems and Equilibrium Models. XVI, 160 pages. 1992.

Vol. 378: W. Krabs, J. Zowe (Eds.). Modern Methods of Optimization. Proceedings, 1990. VIII, 348 pages. 1992.

Vol. 379: K. Marti (Ed.), Stochastic Optimization. Proceedings, 1990. VII, 182 pages. 1992.

Vol. 380: J. Odelstad, Invariance and Structural Dependence. XII, 245 pages. 1992.

Vol. 381: C. Giannini, Topics in Structural VAR Econometrics. XI, 131 pages. 1992.

Vol. 382: W. Oettli. D. Pallaschke (Eds.). Advances in Optimization. Proceedings, 1991. X, 527 pages. 1992.

Vol. 434: M. W. J. Blok, Dynamic Models of the Firm. VII, 193 pages. 1996.

Vol. 435: L. Chen, Interest Rate Dynamics, Derivatives Pricing, and Risk Management. XII, 149 pages. 1996.

Vol. 436: M. Klemisch-Ahlert, Bargaining in Economic and Ethical Environments. IX, 155 pages. 1996.

Vol. 437: C. Jordan, Batching and Scheduling. IX, 178 pages. 1996.

Vol. 438: A. Villar, General Equilibrium with Increasing Returns. XIII, 164 pages. 1996.

Vol. 439: M. Zenner, Learning to Become Rational. VII, 201 pages. 1996.

Vol. 440: W. Ryll, Litigation and Settlement in a Game with Incomplete Information. VIII, 174 pages. 1996.

Vol. 441: H. Dawid, Adaptive Learning by Genetic Algorithms. IX, 166 pages.1996.

Vol. 442: L. Corchón, Theories of Imperfectly Competitive Markets. XIII, 163 pages. 1996.

Vol. 443: G. Lang, On Overlapping Generations Models with Productive Capital. X, 98 pages. 1996.

Vol. 444: S. Jørgensen, G. Zaccour (Eds.), Dynamic Competitive Analysis in Marketing. X, 285 pages. 1996.

Vol. 445: A. H. Christer, S. Osaki, L. C. Thomas (Eds.), Stochastic Modelling in Innovative Manufactoring. X, 361 pages. 1997.

Vol. 446: G. Dhaene, Encompassing. X, 160 pages. 1997.

Vol. 447: A. Artale, Rings in Auctions. X, 172 pages. 1997.

Vol. 448: G. Fandel, T. Gal (Eds.), Multiple Criteria Decision Making. XII, 678 pages. 1997.

Vol. 449: F. Fang, M. Sanglier (Eds.), Complexity and Self-Organization in Social and Economic Systems. IX, 317 pages, 1997.

Vol. 450: P. M. Pardalos, D.W. Hearn. W.W. Hager, (Eds.), Network Optimization. VIII, 485 pages, 1997.

Vol. 451: M. Salge, Rational Bubbles. Theoretical Basis, Economic Relevance, and Empirical Evidence with a Special Emphasis on the German Stock Market.IX, 265 pages. 1997.

Vol. 452: P. Gritzmann, R. Horst, E. Sachs, R. Tichatschke (Eds.), Recent Advances in Optimization. VIII, 379 pages. 1997.

Vol. 453: A. S. Tangian, J. Gruber (Eds.), Constructing Scalar-Valued Objective Functions. VIII, 298 pages. 1997.

Vol. 454: H.-M. Krolzig, Markov-Switching Vector Autoregressions. XIV, 358 pages. 1997.

Vol. 455: R. Caballero, F. Ruiz, R. E. Steuer (Eds.), Advances in Multiple Objective and Goal Programming. VIII, 391 pages. 1997.

Vol. 456: R. Conte, R. Hegselmann, P. Terna (Eds.), Simulating Social Phenomena. VIII, 536 pages. 1997.

Vol. 457: C. Hsu, Volume and the Nonlinear Dynamics of Stock Returns. VIII, 133 pages. 1998.

Vol. 458: K. Marti, P. Kall (Eds.), Stochastic Programming Methods and Technical Applications. X, 437 pages. 1998.

Vol. 459: H. K. Ryu, D. J. Slottje, Measuring Trends in U.S. Income Inequality. XI, 195 pages. 1998.

Vol. 460: B. Fleischmann, J. A. E. E. van Nunen, M. G. Speranza, P. Stähly, Advances in Distribution Logistic. XI, 535 pages. 1998.

Vol. 461: U. Schmidt, Axiomatic Utility Theory under Risk. XV, 201 pages. 1998.

Vol. 462: L. von Auer, Dynamic Preferences, Choice Mechanisms, and Welfare. XII, 226 pages. 1998.

Vol. 463: G. Abraham-Frois (Ed.), Non-Linear Dynamics and Endogenous Cycles. VI, 204 pages. 1998.

Vol. 464: A. Aulin, The Impact of Science on Economic Growth and its Cycles. IX, 204 pages. 1998.

Vol. 465: T. J. Stewart, R. C. van den Honert (Eds.), Trends in Multicriteria Decision Making. X, 448 pages. 1998.

Vol. 466: A. Sadrieh, The Alternating Double Auction Market. VII, 350 pages. 1998.

Vol. 467: H. Hennig-Schmidt, Bargaining in a Video Experiment. Determinants of Boundedly Rational Behavior. XII, 221 pages. 1999.

Vol. 468: A. Ziegler, A Game Theory Analysis of Options. XIV, 145 pages. 1999.

Vol. 469: M. P. Vogel, Environmental Kuznets Curves. XIII, 197 pages. 1999.

Vol. 470: M. Ammann, Pricing Derivative Credit Risk. XII, 228 pages. 1999.

Vol. 471: N. H. M. Wilson (Ed.), Computer-Aided Transit Scheduling. XI, 444 pages. 1999.

Vol. 472: J.-R. Tyran, Money Illusion and Strategic Complementarity as Causes of Monetary Non-Neutrality. X, 228 pages. 1999.

Vol. 473: S. Helber, Performance Analysis of Flow Lines with Non-Linear Flow of Material. IX, 280 pages. 1999.

Vol. 474: U. Schwalbe, The Core of Economies with Asymmetric Information. IX, 141 pages. 1999.

Vol. 475: L. Kaas, Dynamic Macroelectronics with Imperfect Competition. XI, 155 pages. 1999.

Vol. 476: R. Demel, Fiscal Policy, Public Debt and the Term Structure of Interest Rates. X, 279 pages. 1999.

Vol. 477: M. Théra, R. Tichatschke (Eds.), Ill-posed Variational Problems and Regularization Techniques. VIII, 274 pages. 1999.

Vol. 478: S. Hartmann, Project Scheduling under Limited Resources. XII, 221 pages. 1999.

Vol. 479: L. v. Thadden, Money, Inflation, and Capital Formation. IX, 192 pages. 1999.

Vol. 480: M. Grazia Speranza, P. Stähly (Eds.), New Trends in Distribution Logistics. X, 336 pages. 1999.

Vol. 481: V. H. Nguyen, J. J. Strodiot, P. Tossings (Eds.). Optimations. IX, 498 pages. 2000.

Vol. 482: W. B. Zhang, A Theory of International Trade. XI, 192 pages. 2000.

Vol. 483: M. Königstein, Equity, Efficiency and Evolutionary Stability in Bargaining Games with Joint Production. XII, 197 pages. 2000.

GPSR Compliance

The European Union's (EU) General Product Safety Regulation (GPSR) is a set of rules that requires consumer products to be safe and our obligations to ensure this.

If you have any concerns about our products, you can contact us on ProductSafety@springernature.com

In case Publisher is established outside the EU, the EU authorized representative is:

Springer Nature Customer Service Center GmbH
Europaplatz 3
69115 Heidelberg, Germany

Batch number: 10844675

Printed by Printforce, the Netherlands